Toward a PFAS-free Future

Safer Alternatives to Forever Chemicals

Green Chemistry Series

Editor-in-chief:
James H. Clark, *Department of Chemistry, University of York, UK*

Series editors:
Catherine Birch, *AgriFood X Limited, UK*
Graham Bonwick, *AgriFood X Limited, UK*
George A. Kraus, *Iowa State University, USA*
Andrzej Stankiewicz, *Delft University of Technology, The Netherlands*

Titles in the series:
1: The Future of Glycerol: New Uses of a Versatile Raw Material
2: Alternative Solvents for Green Chemistry
3: Eco-Friendly Synthesis of Fine Chemicals
4: Sustainable Solutions for Modern Economies
5: Chemical Reactions and Processes under Flow Conditions
6: Radical Reactions in Aqueous Media
7: Aqueous Microwave Chemistry
8: The Future of Glycerol: 2nd Edition
9: Transportation Biofuels: Novel Pathways for the Production of Ethanol, Biogas and Biodiesel
10: Alternatives to Conventional Food Processing
11: Green Trends in Insect Control
12: A Handbook of Applied Biopolymer Technology: Synthesis, Degradation and Applications
13: Challenges in Green Analytical Chemistry
14: Advanced Oil Crop Biorefineries
15: Enantioselective Homogeneous Supported Catalysis
16: Natural Polymers Volume 1: Composites
17: Natural Polymers Volume 2: Nanocomposites
18: Integrated Forest Biorefineries
19: Sustainable Preparation of Metal Nanoparticles: Methods and Applications
20: Alternative Solvents for Green Chemistry: 2nd Edition
21: Natural Product Extraction: Principles and Applications
22: Element Recovery and Sustainability
23: Green Materials for Sustainable Water Remediation and Treatment
24: The Economic Utilisation of Food Co-Products
25: Biomass for Sustainable Applications: Pollution Remediation and Energy
26: From C–H to C–C Bonds: Cross-Dehydrogenative-Coupling
27: Renewable Resources for Biorefineries
28: Transition Metal Catalysis in Aerobic Alcohol Oxidation
29: Green Materials from Plant Oils

30: Polyhydroxyalkanoates (PHAs) Based Blends, Composites and Nanocomposites
31: Ball Milling Towards Green Synthesis: Applications, Projects, Challenges
32: Porous Carbon Materials from Sustainable Precursors
33: Heterogeneous Catalysis for Today's Challenges: Synthesis, Characterization and Applications
34: Chemical Biotechnology and Bioengineering
35: Microwave-Assisted Polymerization
36: Ionic Liquids in the Biorefinery Concept: Challenges and Perspectives
37: Starch-based Blends, Composites and Nanocomposites
38: Sustainable Catalysis: With Non-endangered Metals, Part 1
39: Sustainable Catalysis: With Non-endangered Metals, Part 2
40: Sustainable Catalysis: Without Metals or Other Endangered Elements, Part 1
41: Sustainable Catalysis: Without Metals or Other Endangered Elements, Part 2
42: Green Photo-active Nanomaterials
43: Commercializing Biobased Products: Opportunities, Challenges, Benefits, and Risks
44: Biomass Sugars for Non-Fuel Applications
45: White Biotechnology for Sustainable Chemistry
46: Green and Sustainable Medicinal Chemistry: Methods, Tools and Strategies for the 21st Century Pharmaceutical Industry
47: Alternative Energy Sources for Green Chemistry
48: High Pressure Technologies in Biomass Conversion
49: Sustainable Solvents: Perspectives from Research, Business and International Policy
50: Fast Pyrolysis of Biomass: Advances in Science and Technology
51: Catalyst-free Organic Synthesis
52: Hazardous Reagent Substitution: A Pharmaceutical Perspective
53: Alternatives to Conventional Food Processing: 2nd Edition
54: Sustainable Synthesis of Pharmaceuticals: Using Transition Metal Complexes as Catalysts
55: Intensification of Biobased Processes
56: Sustainable Catalysis for Biorefineries
57: Supercritical and Other High-pressure Solvent Systems: For Extraction, Reaction and Material Processing
58: Biobased Aerogels: Polysaccharide and Protein-based Materials
59: Rubber Recycling: Challenges and Developments
60: Green Chemistry for Surface Coatings, Inks and Adhesives: Sustainable Applications
61: Green Synthetic Processes and Procedures
62: Resource Recovery from Wastes: Towards a Circular Economy
63: Flow Chemistry: Integrated Approaches for Practical Applications
64: Transition Towards a Sustainable Biobased Economy

65: Transportation Biofuels: Pathways for Production: 2nd Edition
66: Challenges in Green Analytical Chemistry: 2nd Edition
67: CO_2-switchable Materials: Solvents, Surfactants, Solutes and Solids
68: Green Toxicology: Making Chemicals Benign by Design
69: Sustainable and Functional Redox Chemistry
70: Waste Valorisation Using Ionic Liquids
71: Natural Product Extraction: Principles and Applications: 2nd Edition
72: Renewable Resources for Surface Coatings, Inks and Adhesives
73: Green Corrosion Inhibition: Fundamentals, Design, Synthesis and Applications
74: Chemical Valorisation of Carbon Dioxide
75: Green Sample Preparation Techniques: Concepts, Novel Materials and Solvents, and Applications
76: Methane Conversion Routes: Status and Prospects
77: Green Gasoline: A Green Spark Transportation Fuel
78: Agri-food Waste Valorisation
79: Bioplastics and Biocomposites: A Practical Introduction
80: Green Development of Photoluminescent Carbon Dots: For Analytical, Biomedical and Environmental Applications
81: Toward a PFAS-free Future: Safer Alternatives to Forever Chemicals

How to obtain future titles on publication:
A standing order plan is available for this series. A standing order will bring delivery of each new volume immediately on publication.

For further information please contact:
Book Sales Department, Royal Society of Chemistry, Thomas Graham House, Science Park, Milton Road, Cambridge, CB4 0WF, UK
Telephone: +44 (0)1223 420066, Fax: +44 (0)1223 420247
Email: booksales@rsc.org
Visit our website at www.rsc.org/books

Toward a PFAS-free Future
Safer Alternatives to Forever Chemicals

Edited by

Simona A. Bălan
California Department of Toxic Substances Control, USA
Email: simona.balan@dtsc.ca.gov

Thomas A. Bruton
California Department of Toxic Substances Control, USA
Email: tom.bruton@dtsc.ca.gov

and

Kimberly G. Hazard
University of California, Berkeley, USA
Email: hazard@berkeley.edu

ROYAL SOCIETY OF CHEMISTRY

Green Chemistry Series No. 81

Print ISBN: 978-1-83767-054-3
PDF ISBN: 978-1-83767-141-0
EPUB ISBN: 978-1-83767-142-7
Print ISSN: 1757-7039
Electronic ISSN: 1757-7047

A catalogue record for this book is available from the British Library

The Royal Society of Chemistry is a charity, registered in England and Wales, Number 207890, and a company incorporated in England by Royal Charter (Registered No. RC000524), registered office: Burlington House, Piccadilly, London W1J 0BA, UK, Telephone: +44 (0)20 7437 8656.

For further information see our website at www.rsc.org

Printed in the United Kingdom by CPI Group (UK) Ltd, Croydon, CR0 4YY, UK

Foreword

For decades, perfluoroalkyl and polyfluoroalkyl substances (PFASs) have been used in a wide range of consumer and industrial products, from textiles, to food packaging, building materials, and electronics. Most PFASs are environmentally persistent and many have been linked to adverse health outcomes including cancer, reproductive and developmental toxicity, and endocrine disruption (see Chapter 2). This class of chemicals now contaminates nearly 3000 locations in all 50 U.S. states. According to the U.S. Centers for Disease Control and Prevention's biomonitoring program, which measures synthetic chemicals and pollutants in a representative sample of the U.S. population, PFASs are found in the blood of nearly all Americans. Accordingly, PFASs are facing increasing regulation at the state, federal, and international level, and there is a growing demand from businesses, workers, and consumers for safer alternatives.

Several of the case studies in this book originated from an innovative project-based graduate course that we started over a decade ago at University of California, Berkeley. The Greener Solutions course began in partnership with the California Department of Toxic Substances Control to pair interdisciplinary graduate student teams with organizations seeking safer alternatives to hazardous chemicals in products or manufacturing processes. Each fall, teams of students from chemistry, engineering, and the environmental health sciences tackle new challenges posed by partner organizations. Student teams adopt a functional approach to identify safer chemistries and materials, seeking system-level changes that avoid the haphazard substitution of one hazardous chemical for another. The premise of the projects is to elevate human health and environmental impacts as design criteria on par with other measures of technical performance. The results—recommendations for green chemistry and bio-inspired design

Green Chemistry Series No. 81
Toward a PFAS-free Future: Safer Alternatives to Forever Chemicals
Edited by Simona A. Bălan, Thomas A. Bruton and Kimberly G. Hazard

Published by the Royal Society of Chemistry, www.rsc.org

solutions—often simultaneously address additional sustainability goals, such as reduced energy and water use.

Greener Solutions course projects have proposed safer alternatives for preservatives in personal care products, cross-linkers in 3D printing resins, UV blockers in sunscreens, and PFASs in a variety of applications. Projects seeking safer alternatives to PFASs have been a recurring theme in the course, reflecting both the urgency of the health threat these chemicals pose and their ubiquity in the material economy.

The examples in this book provide a practical guide to some existing alternatives to PFASs. Perhaps more importantly, they demonstrate how to broaden the universe of potential solutions by drawing inspiration from biological systems, green chemistry, and circular design thinking. The book also demonstrates how to work in the face of incomplete data to evaluate potential solutions for their technical performance, safety, and sustainability.

Readers will gain fresh insights from this book with new ideas for eliminating PFASs from a wide range of consumer and industrial products. We hope that students, researchers, innovators, and product designers will also be inspired to apply their skills toward designing the next generation of safer chemicals and safer products.

Marty Mulvihill, PhD and Meg Schwarzman, MD, MPH
Berkeley Center for Green Chemistry (BCGC) Board of Directors
and co-founders of BCGC's Greener Solutions course

Preface

This book is a practical reference work for researchers and industrialists seeking to identify safer alternatives to perfluoroalkyl and polyfluoroalkyl substances (PFASs) in various applications. These "forever chemicals" have been widely used in industry and consumer products, raising serious concerns about their impacts on human health and the environment. It is becoming increasingly clear that such problematic chemicals that can last virtually forever in the environment cannot continue to be used forever. A sustainable future cannot be built with unsustainable chemicals.

In this book, we explore the challenges and opportunities in moving toward a PFAS-free future. We examine the science of PFASs, their sources and pathways of exposure, and their impacts on human health and the environment. We also review the regulatory landscape and the emerging policy frameworks aimed at addressing the PFAS crisis.

But this book is not only about the PFAS problem. It is also about developing solutions, offering guidance on how to identify and evaluate safer and more sustainable alternatives. We examine case studies of identifying safer alternatives to PFASs in food and household packaging, textiles, firefighting gear, floor polish, and building materials, as well as options for removing PFASs from carpet fibers during the recycling process. These case studies demonstrate that it is possible to transition away from PFASs while avoiding haphazard substitutions by embracing the principles of green and sustainable chemistry during research and development.

This book is the result of a fruitful and long-standing collaboration between the Safer Consumer Products (SCP) Program at the California Department of Toxic Substances Control (DTSC) and the Berkeley Center for Green Chemistry (BCGC) at the University of California (UC), Berkeley. Both organizations share a common mission of promoting the principles and

Green Chemistry Series No. 81
Toward a PFAS-free Future: Safer Alternatives to Forever Chemicals
Edited by Simona A. Bălan, Thomas A. Bruton and Kimberly G. Hazard

Published by the Royal Society of Chemistry, www.rsc.org

practices of green and sustainable chemistry, and of identifying safer alternatives to hazardous chemicals.

SCP is a California state regulatory program that aims to advance the design, development, and use of products that are chemically safer for people and the environment. The program uses a science-based approach to identify consumer products containing chemicals of concern, and works with product manufacturers to find safer alternatives.

BCGC offers the Greener Solutions graduate course at UC Berkeley, which is a hands-on, project-based course that engages students in the process of identifying and evaluating alternatives to hazardous chemicals in various applications. The course is designed to bridge the gap between academia and industry, and to provide students with the practical skills and knowledge needed to address real-world green chemistry challenges.

This book is a testament to the success of the interdisciplinary collaboration necessary to apply green and sustainable chemistry. Such collaborations will be critical on the road toward a PFAS-free future.

The authors would like to thank Thomas McKeag, former Executive Director of the Berkeley Center for Green Chemistry, for initiating and spearheading this project in the early phases. We also acknowledge the hard work of all the students whose research has produced the proposed solutions in these case studies. We are grateful to David Grealish from DTSC for developing the graphical illustrations in Chapters 1 and 10, and to Helen Armes and Amina Headley from the Royal Society of Chemistry for their publishing support.

Simona A. Bălan
Thomas A. Bruton
Kimberly G. Hazard

Author Biographies

Simona A. Bălan supervises the External Communications and Environmental Justice unit in the Safer Consumer Products Program at the California Department of Toxic Substances Control (DTSC), where she has been leading efforts to regulate the entire class of perfluoroalkyl and polyfluoroalkyl substances (PFASs) in certain consumer products. She also occasionally lectures at the University of California, Berkeley, in the School of Public Health. Before joining DTSC, she was a Senior Scientist at the Green Science Policy Institute, managing international projects on the use of flame retardants and PFASs in consumer products. She holds a PhD in Environmental Science, Policy and Management from UC Berkeley and a BSc in Earth and Planetary Sciences from Jacobs University Bremen, Germany.

Elizabeth Boxer is a PhD student at Duke University in the Integrated Toxicology Environmental Health Program. Prior to joining Duke, she earned her MPH in environmental health sciences at the University of California, Berkeley. She was introduced to PFASs as an Undergraduate Research Assistant at Northeastern University where she helped to create and maintain the first public database of PFAS contamination sites in the U.S. Elizabeth also has experience working as a consultant in the life sciences and transportation industries. Her current research includes studying the reproductive and developmental effects of prenatal PFAS and semi-volatile organic compound (SVOC) exposure.

Marianna Brown-Augustine is a Client Manager at Charles River Laboratories. She works with industrial teams to build pre-clinical research programs using immuno-oncology models for drug development. She received her PhD in molecular toxicology from the University of California, Berkeley.

Green Chemistry Series No. 81
Toward a PFAS-free Future: Safer Alternatives to Forever Chemicals
Edited by Simona A. Bălan, Thomas A. Bruton and Kimberly G. Hazard

Published by the Royal Society of Chemistry, www.rsc.org

Thomas A. Bruton is a Senior Hazardous Substances Engineer at the California Department of Toxic Substances Control, and a former Senior Scientist and Science and Policy Fellow at the Green Science and Policy Institute. He holds a PhD in environmental engineering from UC Berkeley, an MS in civil and environmental engineering from Arizona State University, and a BS in civil engineering from Iowa State University.

Grace Campbell is an MS environmental health sciences student at the University of California, Berkeley. Grace previously earned her BS at UC Berkeley in environmental engineering science, during which she studied the chemical remediation of PFAS in AFFF-contaminated groundwater. After receiving her bachelor's, Grace worked for the NC State Superfund Research Center supporting PFAS research and science communication. She has worked on the issue of PFAS for nearly 4 years, has a diverse knowledge of PFAS research and science, and cares deeply about working towards eliminating PFAS as a class from the entire life cycle.

Emily Cook is an Environmental Engineer at EKI Environment & Water, Inc. where she works on contaminated groundwater and soil investigation and remediation projects in California. Dr Cook earned her doctorate in environmental engineering from the University of California, Berkeley studying the biotransformation of PFAS precursors with various source zone co-contaminants, activated persulfate chemical oxidation and mineralization of PFAS subclasses in the context of *in situ* treatment, and the toxicity of PFAS on *in situ* bioremediation processes. Dr Cook earned her MS in environmental engineering from UC Berkeley and her BS in civil engineering from the University of Kansas.

Erin Creel is currently a Senior Chemical Engineer at a fossil-free hydrogen company. She received a PhD in chemistry from the University of California, Berkeley, studying surface-plasmon-enhanced electrochemical carbon dioxide reduction. Dr Creel earned her BS in chemistry and mathematics from Sewanee: The University of the South.

Edythe Glazer is an MPH student in the School of Public Health, Environmental Health Sciences Department at the University of California, Berkeley. She previously received a BS in Public Health from Tulane University's School of Public Health and Tropical Medicine. Edythe has a broad interest in the health implications of climate change. She is currently working as a Research Assistant in the Sustainability and Health Equity Laboratory at UC Berkeley where she is analyzing occupational and environmental exposure to polycyclic aromatic hydrocarbons (PAHs) among a cohort of female firefighters at the San Francisco Fire Department.

Kelly A. Grant is a Senior Environmental Scientist at the California Department of Toxic Substances Control, where she works in the Safer Consumer Products Program (SCP), regulating chemicals in products. Prior to joining DTSC, Kelly served as an American Association for the Advancement of

Science (AAAS) Policy Fellow in U.S. EPA's Design for the Environment. She received her PhD in molecular and cellular biology from the University of Washington, Seattle and a BS in molecular cellular and developmental biology from Haverford College in Pennsylvania.

Amanda Guan is an MS student in the Materials Science and Engineering department at the University of California, Berkeley. As an undergraduate student at Berkeley, she investigated safer, biopolymer alternatives to PFAS and polyolefin films in product packaging through the Greener Solutions course. This project ultimately provided the foundation for her research thesis in collaboration with Sway, where she studies and optimizes seaweed-based replacements to oil-based, single-use packaging. Previously, Amanda has also developed high-performance 3D printing resins, qualified polymer systems in aerospace and defense, and led a team to lobby for environmental protection in Washington, DC.

Kimberly G. Hazard is a PhD candidate in the School of Public Health, Environmental Health Sciences division, University of California, Berkeley. She is the teaching and outreach program manager for the Berkeley Center for Green Chemistry and a co-instructor of the graduate course, Greener Solutions. She holds a MS in environmental health sciences and a BS in society and environment from UC Berkeley. Her research focuses on chemical exposures in early childhood environments and her teaching focuses on chemical hazard assessment.

Christopher Leonetti was a Senior Environmental Scientist at the California Department of Toxic Substances Control (DTSC) from 2018–2023, where he worked in the Safer Consumer Products (SCP) Program. He worked on projects aimed at regulating the entire class of PFASs within various consumer product categories. He holds a PhD in integrated toxicology and environmental health from Duke University and a BS in biology from the University of Michigan – Ann Arbor.

Aaron Maruzzo is an MPH student in the Environmental Health Science Department at UC Berkeley School of Public Health. He was previously a Water Lab Analyst for the municipal water company in the U.S. territory, the Commonwealth of the Northern Marianas Islands, serving the islands of Saipan, Tinian, and Rota. His current research and advocacy focus on PFAS, which have been detected in the water supplies in Guam and Saipan. Incorporating an interdisciplinary approach, his work aims to mitigate toxic exposures and move toward environmental justice through community-based action.

Sumana Raj is currently a Research Associate at the SLAC National Accelerator Laboratory. She received her PhD in physical chemistry from the University of California, Berkeley, and a BA in chemistry from Cornell University.

Hannah Ray, PhD, is the Senior Manager, Programs + Innovation (Materials) at the International Living Future Institute, where she develops and manages the Red List, the Materials Petal of the Living Building Challenge, the Declare program, and the Living Product Challenge. She has professional experience in research and development at an electrochromic glass manufacturer, and advocacy and material health research experience from the Green Science Policy Institute. Hannah received her PhD in materials science from UC Berkeley, and her BA in chemistry from Wesleyan University.

Brittany Stinger is a chemistry PhD candidate in the Jay Groves lab at UC Berkeley with an expected graduation in May 2025. She received her BS in chemistry from Rowan University. Her current research focuses on understanding the role of various proteins involved in the T-cell signaling pathway. Brittany is interested in various aspects of green chemistry, specifically related to the elimination/minimization of waste and the design of safer chemicals.

Jenna Tan received a PhD in chemistry from the University of California, Berkeley. Her research focused on characterizing the crystallization of organic semiconductor molecules for thin-film and semiconductor technologies. She employed her experience in evaluating molecular physicochemical properties toward understanding the function of PFAS in floor polish formulations and developing the technical metrics for assessing PFAS alternatives.

Minerva Teli is an environmental engineering PhD student at the University of California, Berkeley. She has a background in chemistry and materials science, which she uses to develop water treatment and sensing technologies. She applies this same background to her work in finding PFAS alternatives, which began with Berkeley's Greener Solutions graduate course, followed by her work with the Washington State Department of Ecology. As part of Ecology's Reducing Toxic Threats team, Minerva contributed to the research and writing of an alternatives assessment that will effectively ban the use of PFAS from food packaging within the Washington State. More broadly, her interests include green chemistry, sustainable technology development, and environmental justice.

McKenna Thompson is an MPH student in Environmental Health Sciences at the University of California, Berkeley. Her research focuses on characterizing and mitigating exposures to air pollutants in heavily burdened communities. She has a background in advocacy, policy analyses and epidemiology. She holds a BS degree in public health from the University of California, San Diego.

Sophia Thompson is an MPH environmental health sciences 2023 graduate from the University of California, Berkeley, with a focus on global health and environment. She has a professional background in environmental health and safety at UC Santa Barbara along with an academic background in policy review and analysis.

Tessa Wardle received her MS degree in environmental health sciences from the UC Berkeley School of Public Health. Her experience lies in assessing the health consequences of toxic chemicals, and she has worked on multiple projects relating to the current crisis of widespread PFAS contamination. In the past, she has worked with organizations advocating for environmental justice, encouraging conservation and smart growth in the Bay Area, reducing farmworker communities' exposure to pesticides in California, and researching air releases from industrial facilities in the United States.

Yuning Xu is an MPH student in the Environmental Health Sciences Program at the University of California, Berkeley, with a focus on industrial hygiene. She has experience of working on a project related to PFAS policy in waste water while applying the p-sufficient approach. Yuning used her background to evaluate the health and environmental endpoints of PFAS and the alternatives under a life cycle analysis framework and researched studies for the technical performance of PFAS alternatives.

Contents

Green Chemistry Series No. 81
Toward a PFAS-free Future: Safer Alternatives to Forever Chemicals
Edited by Simona A. Bălan, Thomas A. Bruton and Kimberly G. Hazard

Published by the Royal Society of Chemistry, www.rsc.org

CHAPTER 1

Why Aim Toward a PFAS-free Future?

S. A. BĂLAN,*[a,b] T. A. BRUTON[a] AND K. GRANT[a]

[a] California Department of Toxic Substances Control, Sacramento, CA, 95812, USA; [b] University of California, Berkeley, School of Public Health, Berkeley, CA, 94704, USA
*Email: simona.balan@dtsc.ca.gov

1.1 Introduction

As global consumption continues to grow and society runs up against planetary boundaries,[1] there has never been a greater need for green chemistry. The intersecting global crises of climate change, pollution, and biodiversity loss highlight the need for safe- and sustainable-by-design chemicals and materials[2] that can enable the decarbonization of energy systems and the transition to a circular economy while minimizing the adverse impacts on human and ecological health. A sustainable future cannot be built with unsustainable chemicals.

Perfluoroalkyl and polyfluoroalkyl substances (PFASs) constitute a large class of synthetic chemicals that have become widespread environmental pollutants. They have been detected in indoor and outdoor environments, humans, and biota, including remote mountaintops,[3] the deep sea,[4,5] rainwater,[6] and umbilical cord blood.[7–9] They have even been called "forever chemicals" since they will likely outlast humanity. But despite their widely documented hazards (see Chapter 2), PFASs are still used in hundreds of consumer and industrial products[10,11] due to their unique and lucrative properties. It is difficult to name a manufacturing sector that has not benefited from the use of PFASs. These chemicals have become ingrained in

Green Chemistry Series No. 81
Toward a PFAS-free Future: Safer Alternatives to Forever Chemicals
Edited by Simona A. Bălan, Thomas A. Bruton and Kimberly G. Hazard

Published by the Royal Society of Chemistry, www.rsc.org

a wide range of industries and applications, including surface treatments for textiles and food packaging, pharmaceuticals, pesticides, aqueous film-forming firefighting foams (AFFFs), metal plating fume suppressants, aerospace components, automotive parts and interiors, paints and other architectural coatings, flooring, roofing, semiconductors, touchscreens, lithium-ion batteries, and many more.[10–14]

Depending on whom you ask, a PFAS-free future might seem both unattainable and inevitable. On one hand, scientists around the world have been raising concerns and advocating for the regulation of the entire class of PFASs,[15–20] policymakers have been banning PFASs from a growing list of non-essential uses, consumers are increasingly demanding PFAS-free products, and leading companies are finding innovative safer replacements. On the other hand, the use of PFASs is still mandated by government standards such as the United States (U.S.) Navy's specification for AFFF[21] and the Japanese National Specification for Steel Bridge Coatings.[22] Moreover, the use of PFASs continues to grow in applications promoted for climate change mitigation, such as electric vehicle batteries, solar panels, and wind turbines.

We are now at an inflection point, where policymakers and the public are becoming increasingly aware of the dangers associated with continued PFAS use, triggering widespread changes in the market. This represents an opportunity for businesses of all sizes who incorporate the principles of green and sustainable chemistry into their research and development to emerge as leaders. Replacing PFASs in their diverse applications will require creative thinking and perseverance. There are no prior examples where efforts have been made to find safer alternatives for such a large class of chemicals across such a wide range of products. This transition presents both challenges and opportunities.

To aid in this effort, this book showcases case studies of identifying safer substitutes for PFASs in select consumer products. Most of these case studies were conducted by students of the Greener Solutions course at the University of California Berkeley. The framework employed in these case studies can be applied to other settings and products, with a few modifications.

This chapter provides key definitions and background information on PFASs and their many functional applications. Chapter 2 delves deeper into why PFASs are chemicals of concern, highlighting their health and environmental hazards as well as current and proposed policies at the state, federal, and international levels to address the use of PFASs in consumer products. Chapters 3 through 5 present case studies for identifying safer alternatives to PFASs for food packaging, household packaging, and floor polishes, respectively. Chapter 6 discusses strategies for removing PFASs from carpet fibers during the recycling process, and Chapter 7 addresses how the green building industry is searching for safer alternatives to PFASs for building products. Chapters 8 and 9 present case studies for identifying safer alternatives to PFASs in textiles, for outdoor gear and firefighting turnout gear, respectively. Chapter 10 examines the key factors that are powering the shift toward a PFAS-free future and concludes with recommendations for those willing to embark on this journey.

1.2 What Are PFASs?

In a landmark 2011 paper, Buck *et al.* first introduced the term perfluoroalkyl and polyfluoroalkyl substances (PFASs), differentiating these chemicals from other members of the broader fluorochemicals family.[23] According to their definition, perfluoroalkyl substances are "aliphatic substances for which all of the H atoms attached to C atoms in the non-fluorinated substance from which they are notionally derived have been replaced by F atoms, except for those H atoms whose substitution would modify the nature of any functional groups present", and polyfluoroalkyl substances are "aliphatic substances in which all H atoms attached to at least one (but not all) C atom have been replaced by F atoms, in such a manner that they contain the perfluoroalkyl moiety C_nF_{2n+1}".[23] This definition covers a wide range of substances that vary in terms of their chemical structure and molecular weight.

In 2015, the Swedish Chemicals Agency published a list of over 3000 chemicals that align with the Buck *et al.* PFAS definition.[12] Subsequently, various organizations and authoritative bodies have developed their own working definition and understanding of this class. For instance, in 2018 the Organisation for Economic Cooperation and Development (OECD) published a list of 4730 Chemical Abstracts Service (CAS) registry numbers related to individual PFASs or commercial PFAS mixtures available on the global market that "contain a perfluoroalkyl moiety with three or more carbons (*i.e.*, $-C_nF_{2n}-$, $n \geq 3$) or a perfluoroalkyl ether moiety with two or more carbons (*i.e.*, $-C_nF_{2n}OC_mF_{2m}-$, n and $m \geq 1$)".[24] In 2021, OECD proposed the following revised definition to address gaps in the original Buck *et al.* definition: "PFASs are defined as fluorinated substances that contain at least one fully fluorinated methyl or methylene carbon atom (without any H/Cl/Br/I atom attached to it), *i.e.*, with a few noted exceptions, any chemical with at least a perfluorinated methyl group ($-CF_3$) or a perfluorinated methylene group ($-CF_2-$) is a PFAS".[25]

Meanwhile, the U.S. Environmental Protection Agency (EPA)'s Office of Pollution Prevention and Toxics, which administers the U.S. Toxic Substances Control Act, adopted a more restrictive PFAS working definition that "includes perfluorinated and polyfluorinated substances that structurally contain a $R-CF_2-C(F)(R')(R'')$ unit. Both the CF_2 and CF moieties are saturated and none of the R groups (R, R′ or R″) can be hydrogen".[26] In contrast, several laws in U.S. states limiting the presence of PFASs in certain consumer products broadly define PFASs as "a class of fluorinated organic chemicals containing at least one fully fluorinated carbon atom".[27–35] Some community health advocates are pushing for an even broader definition that would encompass all fluorinated organic chemicals as PFASs, while some industry representatives are arguing that fluoropolymers should not be considered PFASs because they align with the definition of polymers of low concern as outlined by the OECD.[14,36]

Regardless of the specifics of the definition used, a few key subclasses of PFASs can be identified: perfluoroalkyl acids (PFAAs), PFAA precursors, perfluoropolyethers (PFPEs), and fluoropolymers. While PFAAs constitute a small subset (approximately 1%) of PFASs, they are terminal degradation products, manufacturing aids, feedstocks, or impurities of other PFAS class members, which makes their hazard traits relevant to the entire class (Figure 1.1). PFAAs are extremely persistent and mobile in the environment, and the ones which have been studied were found to be toxic (see Chapter 2). This subgroup includes:

- Perfluoroalkyl carboxylic acids (PFCAs) such as perfluorooctanoate (PFOA);[†]
- Perfluoroalkane sulfonic acids (PFSAs) such as perfluorooctane sulfonate (PFOS)[†]
- Perfluoroalkane sulfinic acids (PFSiAs)
- Perfluoroalkyl phosphonic acids (PFPAs)
- Perfluoroalkyl phosphinic acids (PFPiAs)
- Perfluoroether carboxylic acids (PFECAs)
- Perfluoroether sulfonic acids (PFESAs).

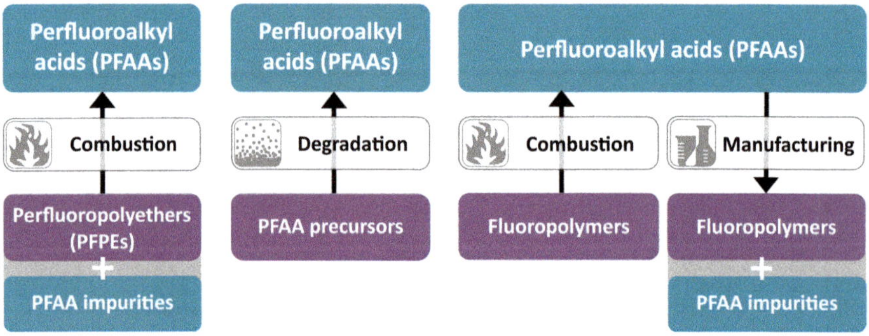

Figure 1.1 Use of all PFAS subclasses can lead to exposure to PFAAs at some point in the chemicals' life cycle; they either degrade to form PFAAs, release PFAAs if incinerated, or require PFAAs for their manufacture, often resulting in PFAA impurities in the final product. Specifically, fluoropolymers are made using PFAAs; PFPEs and fluoropolymers can contain PFAA impurities; PFAAs can be released from PFPEs and fluoropolymers during combustion, and PFAA precursors such as fluorotelomer-based substances, including side-chain fluorinated polymers, degrade to PFAAs. First published in ref. 80. Reproduced with permission from the State of California.

[†]PFOA and PFOS are referred to as "C8" because they contain eight carbon atoms in their molecules.

PFASs that can degrade into PFAAs are termed PFAA precursors and provide a significant indirect source of PFAAs to the environment (see examples of PFAA precursors in Figure 1.2). Most polyfluoroalkyl substances are PFAA precursors. Of the 4730 individual PFASs or commercial PFAS mixtures identified by OECD on the global market, 4186 likely degrade to PFAAs in the environment or biota.[24] Degradation to PFAAs can occur, for instance, in the atmosphere, wastewater treatment plants, landfills, or consumer products, mainly *via* biodegradation, hydrolysis, or photo-oxidation.[37–39] The degradation half-life of PFAA precursors can range from days to centuries.[37,40,41] PFAAs and some of their precursors are frequently subdivided into longer- and shorter-chain compounds. The longer-chain PFSAs have six or more perfluorinated carbons; longer-chain PFCAs, PFPAs, and PFPiAs have seven or more perfluorinated carbons.

PFPEs and fluoropolymers are characterized by fluorine atoms embedded in the backbone of a polymeric structure. PFPEs are perfluoroalkyl substances with large molecular weight (oligomers, polymers, and copolymers) and ether linkages. They are highly persistent in the environment and unlikely to degrade to PFAAs under typical environmental conditions, but may contain non-polymeric impurities and may release hazardous products during combustion.[42] Fluoropolymers are characterized by F atoms directly attached to a carbon-only chain, and are also highly persistent in the environment. They do not degrade to PFAAs under typical environmental

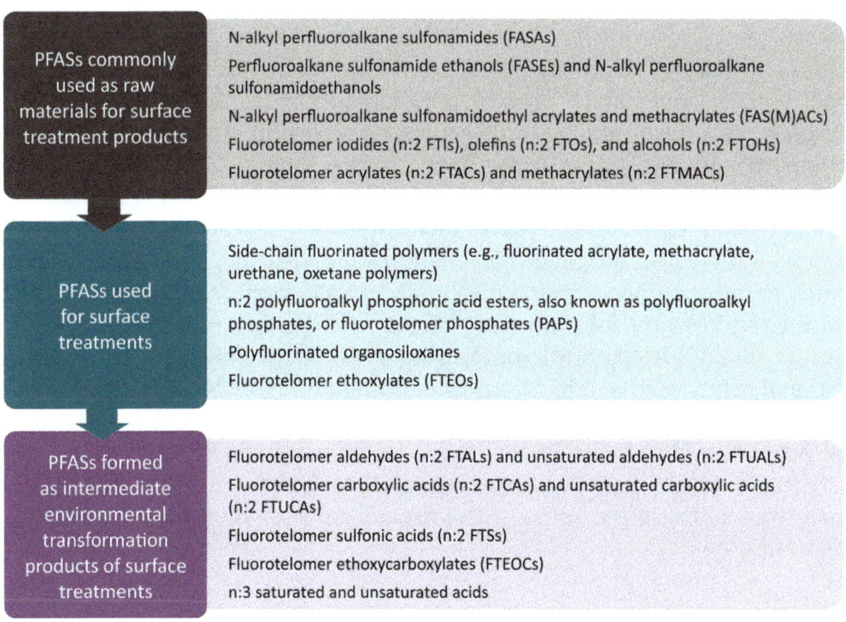

Figure 1.2 Examples of PFAA precursors, as described in Buck *et al.*[23] First published in ref. 80. Reproduced with permission from the State of California.

conditions, but will eventually break down into microplastics by weathering and physical stress, enabling further dispersion and enhanced bioavailability.[43] PFAA manufacturing aids and other low molecular weight intermediaries can occur as impurities in the final fluoropolymer product.[43] Moreover, fluoropolymers may release PFCAs, including PFOA, during combustion at temperatures between 180 and 800 °C.[44,45] Examples of fluoropolymers include polytetrafluoroethylene (PTFE), polyvinylidene fluoride (PVDF), and polyvinyl fluoride (PVF).

1.3 What Makes PFASs Unique?

Fluorine (F) stands out among the elements in the periodic table because it forms the strongest single bond with carbon.[46] Due to its high electronegativity, the F atom pulls the shared electrons of the covalent C–F bond toward itself. This renders it partially negative and the C atom partially positive, creating a polar covalent bond. This bond requires high energy to dissociate, up to 544 kJ mol^{-1},[47] compared to the 337 kJ mol^{-1} needed to separate carbon and hydrogen.[48] This means PFASs, which have multiple carbon–fluorine bonds, tend to be thermally, chemically, and biologically stable.[49] The PFASs that can degrade in the environment typically transform into smaller PFASs that do not degrade further under environmental conditions, such as PFAAs.[23,41,50–53]

When added to products and materials, PFASs confer increased stability at extreme temperatures, antistatic properties, and resistance to wettability, staining, and corrosion.[23,54] Some PFASs are inert to solvents, hydrocarbons, acids, and bases.[55] Of note, PFASs are the only chemicals known to date to repel both oil and water.[56] The oleophobic and hydrophobic characteristics of PFASs are a result of the bonds between the carbon–fluorine (CF_3 and CF_2) groups, which create very low critical surface energy. Their oleophobicity, hydrophobicity, and ability to lower surface tension makes PFASs useful in a variety of materials and products, such as textiles, leathers, food packaging, photovoltaic cells, architectural membranes, paints, and glass.[10] Their ability to lower surface tension also makes PFASs useful in applications such as toner and printer inks, polymer extrusion aids, metal plating, oil and gas drilling fluids, photographic processing solutions, semiconductor industry, textile dyeing, paints and coatings, and AFFF.[10] Glüge *et al.* provided a comprehensive overview of the various industries that employ PFASs, and the properties that make PFASs useful in those applications.[10]

These versatile chemicals have a plethora of functional applications. For instance, these are some of the functional uses identified for PFASs in cosmetics:[12,57–62]

- Adsorbent
- Anticaking
- Antistatic
- Binding

- Brightening
- Bulking
- Cleansing
- Detacktifier
- Detangling
- Durability
- Emollient
- Emulsifying
- Emulsion stabilizing
- Epilating agent
- Film forming
- Hair conditioning
- Hair lamination
- Increasing oxygen absorption of skin
- Increasing skin penetration
- Oil repellency
- Opacifying
- Pore strip
- Protectant
- Skin conditioning
- Skin feel modifier
- Slip modifier
- Solvent
- Surface modifier
- Surfactant
- Surfactant dispersing
- Viscosity controlling
- Water repellency
- Weather resistance

The unique properties of PFASs contribute to their environmental persistence, global distribution, and accumulation in water, soils, plants, and animals. In addition to being persistent, some PFASs are known to be toxic, and some are also bioaccumulative. However, unlike other persistent, bioaccumulative, and toxic contaminants, these and many other PFASs are highly water soluble,[63] which means that drinking water can be a significant route of human exposure to these compounds.

1.4 Avoiding Regrettable Substitutions

Many of the problematic uses of PFASs have resulted from attempts to solve other problems. In many cases, today's problems are yesterday's solutions. In the field of chemical management, the term "regrettable substitution" describes instances where a chemical of known concern is phased out and replaced with a similar chemical with little or no data about its safety. Such replacement chemicals, after years of use in consumer products or industrial applications, are often found to be just as bad, or even worse, than the chemicals they replaced.

An example of regrettable substitution is the case of AFFF. Originally developed by the U.S. Navy and 3M as a way to quickly extinguish liquid fuel fires on ships, this product has been used widely at military installations, airports, and industrial facilities.[64] The surface-active properties of PFASs make them especially suited for use in AFFF, as they allow the product to form a film that spreads on top of the fuel and limits oxygen flow to the fire.[65]

From the 1960s until the early 2000s, the majority of AFFF formulations contained long-chain PFASs.[66] When concerns emerged regarding the adverse impacts of long-chain PFASs, AFFF manufacturers began replacing them with short-chain PFASs that were presumed to have an improved health and safety profile. In the two decades since this switch occurred, further research has shown that the short-chain PFASs used in AFFF are also persistent and can be toxic. Like their long-chain predecessors, use of these chemicals in AFFF contributes to the contamination of water resources.[67]

Governments around the world are now restricting the use of AFFF containing any PFASs. AFFF users who only recently switched from formulations based on long-chain PFASs to those based on short-chain PFASs must now replace their inventories once again and find a method to safely dispose of these foams. The increased regulatory scrutiny on firefighting foams has spurred the development of new PFAS-free formulations and even new frameworks for assessing the environmental profile of these products.[68]

In contrast to regrettable substitutions, informed substitution considers the inherent trade-offs that are involved in replacing a chemical of concern with another chemical or redesigning a product to provide the same function without the need of a chemical. Alternatives assessment is a tool that provides a structured framework for identifying these trade-offs and managing them.

Understanding and mitigating the trade-offs of new products, processes, and technologies are critical for a truly sustainable PFAS-free future. For example, molded fiber food packaging such as clamshell takeout food containers, food bowls, and egg trays was developed as a sustainable alternative to plastic food packaging. Molded fibers are typically derived from agricultural waste products, thus enhancing their value and reducing waste. Moreover, molded fiber products can be composted at the end-of-their-life unlike most plastic packaging, which is typically landfilled. However, molded fibers are not inherently grease- and liquid-proof. To enable this packaging to hold foods such as soup, french fries, salads, *etc.*, PFASs are mixed into the molded fiber slurry during manufacturing. Several studies found PFASs, including PFAAs and their precursors, in molded fiber food packaging products.[69–71] The problem? These PFASs can migrate from packaging products into food,[69,72] resulting in human exposures. Additionally, the presence of highly persistent PFASs conflicts with product compostability claims. Compost derived from PFAS-containing food packaging contains PFASs,[73] which can be taken up by plants from the soil[74–76] and subsequently enter the food chain, impacting both humans and other biota.

Back in 2018, all molded fiber food packaging products on the U.S. market contained PFASs.[77] Since then, pressure from customers, compostability certifiers, and policymakers has motivated some manufacturers to develop PFAS-free molded fiber food packaging products, demonstrating that it is possible to find functionally acceptable alternatives. Based on an alternatives assessment, Washington State concluded that some of these alternatives are safer than PFASs.[78,79] However, these alternatives include coating the molded fiber with a layer of plastic or replacing the molded fiber altogether with certain types of plastic. Most of these plastics are not compostable, so they negate one of the main purported benefits of molded fiber food packaging, *i.e.*, its compostability. Trade-offs such as these are common. Exploring them through an alternatives assessment can guide innovation toward truly sustainable alternatives.

We will revisit this example in more detail in Chapter 3, which provides a framework for identifying safer alternatives to PFASs in plant fiber-based food packaging products. This and the other case studies presented in this book illustrate the benefits of understanding chemical hazards and seeking inspiration from nature for bio-inspired designs. Nature's infinite creativity may be the ultimate ally in the quest toward a PFAS-free future.

Acknowledgements

The authors would like to thank David Grealish for developing the graphical illustrations, and Anne Cooper Doherty and Chris Leonetti for their helpful comments on the manuscript.

References

1. J. Rockström, W. Steffen, K. Noone, Å. Persson, F. S. I. Chapin, E. Lambin, T. M. Lenton, M. Scheffer, C. Folke, H. J. Schellnhuber, B. Nykvist, C. A. de Wit, T. Hughes, S. van der Leeuw, H. Rodhe, S. Sörlin, P. K. Snyder, R. Costanza, U. Svedin, M. Falkenmark, L. Karlberg, R. W. Corell, V. J. Fabry, J. Hansen, B. Walker, D. Liverman, K. Richardson, P. Crutzen and J. Foley, *Ecol. Soc.*, 2009, **14**, 32.
2. European Commission, Chemicals Strategy for Sustainability Towards a Toxic-Free Environment, 2020, https://eur-lex.europa.eu/LexUriServ/LexUriServ.do?uri=SWD:2020:0250:FIN:EN:PDF.
3. K. R. Miner, H. Clifford, T. Taruscio, M. Potocki, G. Solomon, M. Ritari, I. E. Napper, A. P. Gajurel and P. A. Mayewski, *Sci. Total Environ.*, 2021, **759**, 144421.
4. A. Sanchez-Vidal, M. Llorca, M. Farré, M. Canals, D. Barceló, P. Puig and A. Calafat, *Sci. Total Environ.*, 2015, **526**, 41–48.
5. Z. Zhao, Z. Xie, A. Möller, R. Sturm, J. Tang, G. Zhang and R. Ebinghaus, *Environ. Pollut.*, 2012, **170**, 71–77.

6. I. T. Cousins, J. H. Johansson, M. E. Salter, B. Sha and M. Scheringer, *Environ. Sci. Technol.*, 2022, **56**(16), 11172–11179.
7. R. Monroy, K. Morrison, K. Teo, S. Atkinson, C. Kubwabo, B. Stewart and W. G. Foster, *Environ. Res.*, 2008, **108**, 56–62.
8. W. Cao, X. Liu, X. Liu, Y. Zhou, X. Zhang, H. Tian, J. Wang, S. Feng, Y. Wu, P. Bhatti, S. Wen and X. Sun, *Environ. Int.*, 2018, **116**, 197–205.
9. D. Cai, Q.-Q. Li, C. Chu, S.-Z. Wang, Y.-T. Tang, A. A. Appleton, R.-L. Qiu, B.-Y. Yang, L.-W. Hu, G.-H. Dong and X.-W. Zeng, *Sci. Total Environ.*, 2020, **705**, 135885.
10. J. Glüge, M. Scheringer, I. T. Cousins, J. C. DeWitt, G. Goldenman, D. Herzke, R. Lohmann, C. A. Ng, X. Trier and Z. Wang, *Environ. Sci.: Processes Impacts*, 2020, **22**, 2345–2373.
11. L. G. T. Gaines, *Am. J. Ind. Med.*, 2023, **66**(5), 353–378.
12. KEMI, *Occurrence and use of highly fluorinated substances and alternatives, Report from a government assignment, Swedish Chemicals Agency (KEMI)*, Stockholm, Sweden, 2015.
13. American Chemistry Council, Fluorotechnology makes important products for vital industries possible, https://www.americanchemistry.com/Fluorotechnology-Starburst-Uses-Graphic.pdf (accessed 28 January 2021).
14. S. H. Korzeniowski, R. C. Buck, R. M. Newkold, A. E. Kassmi, E. Laganis, Y. Matsuoka, B. Dinelli, S. Beauchet, F. Adamsky, K. Weilandt, V. K. Soni, D. Kapoor, P. Gunasekar, M. Malvasi, G. Brinati and S. Musio, *Integr. Environ. Assess. Manage.*, 2023, **19**(2), 326–354.
15. M. Scheringer, X. Trier, I. T. Cousins, P. de Voogt, T. Fletcher, Z. Wang and T. F. Webster, *Chemosphere*, 2014, **114**, 337–339.
16. A. Blum, S. A. Balan, M. Scheringer, X. Trier, G. Goldenman, I. T. Cousins, M. Diamond, T. Fletcher, C. Higgins, A. E. Lindeman, G. Peaslee, P. de Voogt, Z. Wang and R. Weber, *Environ. Health Perspect.*, 2015, **123**, A107–A111.
17. A. Ritscher, Z. Wang, M. Scheringer, J. M. Boucher, L. Ahrens, U. Berger, S. Bintein, S. K. Bopp, D. Borg, A. M. Buser, I. Cousins, J. DeWitt, T. Fletcher, C. Green, D. Herzke, C. Higgins, J. Huang, H. Hung, T. Knepper, C. S. Lau, E. Leinala, A. B. Lindstrom, J. Liu, M. Miller, K. Ohno, N. Perkola, Y. Shi, L. Småstuen Haug, X. Trier, S. Valsecchi, K. van der Jagt and L. Vierke, *Environ. Health Perspect.*, 2018, **126**, 084502.
18. I. T. Cousins, J. C. DeWitt, J. Glüge, G. Goldenman, D. Herzke, R. Lohmann, C. A. Ng, M. Scheringer and Z. Wang, *Environ. Sci.: Processes Impacts*, 2020, **22**(12), 2307–2312.
19. C. F. Kwiatkowski, D. Q. Andrews, L. S. Birnbaum, T. A. Bruton, J. C. DeWitt, D. R. U. Knappe, M. V. Maffini, M. F. Miller, K. E. Pelch, A. Reade, A. Soehl, X. Trier, M. Venier, C. C. Wagner, Z. Wang and A. Blum, *Environ. Sci. Technol. Lett.*, 2020, **7**(8), 532–543.
20. S. A. Bălan, V. C. Mathrani, D. F. Guo and A. M. Algazi, *Environ. Health Perspect.*, 2021, **129**, 025001.
21. U.S. Department of the Navy, MIL-F-24385F (Amendment 1), Military Specification: Fire Extinguishing Agent, Aqueous Film Forming Foam (AFFF)

Liquid Concentrate, for Fresh and Seawater, 1994, http://everyspec.com/MIL-SPECS/MIL-SPECS-MIL-F/MIL-F-24385F_AMENDMENT-1_9881/.

22. Lumiflon FEVE Resins, The Quest For 100-Year Bridge Coatings In Japan – Lumiflon FEVE Resins, https://lumiflonusa.com/technical/the-quest-for-100-year-bridge-coatings-in-japan/ (accessed 1 July 2022).
23. R. C. Buck, J. Franklin, U. Berger, J. M. Conder, I. T. Cousins, P. de Voogt, A. A. Jensen, K. Kannan, S. A. Mabury and S. P. van Leeuwen, *Integr. Environ. Assess. Manage.*, 2011, 7, 513–541.
24. OECD, *Toward a new comprehensive global database of per and polyfluoroalkyl substances (PFASs): Summary report on updating the OECD 2007 list of per- and polyfluoroalkyl substances (PFASs)*, Organisation for Economic Cooperation and Development (OECD), 2018.
25. OECD, *Reconciling terminology of the universe of per- and polyfluoroalkyl substances: Recommendations and practical guidance*, Organisation for Economic Co-operation and Development (OECD), 2021.
26. U.S. EPA, TSCA Section 8(a)(7) Reporting and Recordkeeping Requirements for Perfluoroalkyl and Polyfluoroalkyl Substances, https://www.federalregister.gov/documents/2021/06/28/2021-13180/tsca-section-8a7-reporting-and-recordkeeping-requirements-for-perfluoroalkyl-and-polyfluoroalkyl (accessed 26 May 2022).
27. Washington State Legislature, Engrossed Substitute Senate Bill 6413, 2018.
28. Washington State Legislature, Engrossed Substitute House Bill 2658, 2018.
29. California Legislature, Use of Fire Equipment. Health and Safety Code, Division 12, Part 1, Chapter 2, Article 2, https://leginfo.legislature.ca.gov/faces/codes_displaySection.xhtml?sectionNum=13061.&nodeTreePath=19.1.2.2&lawCode=HSC (accessed 31 May 2022).
30. New York State, NY State Senate Bill S7167, https://www.nysenate.gov/legislation/bills/2019/s7167 (accessed 2 June 2022).
31. New York State, NY State Senate Bill S8817, https://www.nysenate.gov/legislation/bills/2019/s8817 (accessed 2 June 2022).
32. California Legislature, Juvenile Products. Health and Safety Code, Division 104, Part 3, Chapter 12.5, https://leginfo.legislature.ca.gov/faces/codes_displaySection.xhtml?sectionNum=108945.&nodeTreePath=43.3.13&lawCode=HSC (accessed 31 May 2022).
33. California Legislature, Plant-Based Food Packaging Containing PFAS. Health and Safety Code, Division 104, Part 3, Chapter 15, Article 1, https://leginfo.legislature.ca.gov/faces/codes_displaySection.xhtml?lawCode=HSC&sectionNum=109000.&article=1.&highlight=true&keyword=PFAS (accessed 31 May 2022).
34. Vermont Legislature, S.20 (Act 36), An act relating to restrictions on perfluoroalkyl and polyfluoroalkyl substances and other chemicals of concern in consumer products, https://legislature.vermont.gov/bill/status/2022/S.20 (accessed 20 May 2021).

35. State of Maine, An Act To Stop Perfluoroalkyl and Polyfluoroalkyl Substances Pollution, http://www.mainelegislature.org/legis/bills/getPDF.asp?paper=HP1113&item=5&snum=130 (accessed 2 June 2022).
36. B. J. Henry, J. P. Carlin, J. A. Hammerschmidt, R. C. Buck, L. W. Buxton, H. Fiedler, J. Seed and O. Hernandez, *Integr. Environ. Assess. Manage.*, 2018, **14**, 316–334.
37. C. J. Nielsen, *Potential PFOA Precursors. Literature study and theoretical assessment of abiotic degradation pathways leading to PFOA, M-231/2014*, University of Oslo, 2014.
38. K. Prevedouros, I. T. Cousins, R. C. Buck and S. H. Korzeniowski, *Environ. Sci. Technol.*, 2006, **40**, 32–44.
39. H. Hamid, L. Y. Li and J. R. Grace, *Environ. Pollut.*, 2020, **259**, 113835.
40. L. Li, J. Liu, J. Hu and F. Wania, *Environ. Sci. Technol.*, 2017, **51**, 4461–4470.
41. J. W. Washington, T. M. Jenkins, K. Rankin and J. E. Naile, *Environ. Sci. Technol.*, 2015, **49**, 915–923.
42. Nordic Council of Ministers, *Per- and polyfluoroalkylether substances: Identity, production and use*, 2020.
43. R. Lohmann, I. T. Cousins, J. C. DeWitt, J. Glüge, G. Goldenman, D. Herzke, A. B. Lindstrom, M. F. Miller, C. A. Ng, S. Patton, M. Scheringer, X. Trier and Z. Wang, *Environ. Sci. Technol.*, 2020, **54**, 12820–12828.
44. M. Feng, R. Qu, Z. Wei, L. Wang, P. Sun and Z. Wang, *Sci. Rep.*, 2015, **5**(1), 9859.
45. M. Schlummer, C. Sölch, T. Meisel, M. Still, L. Gruber and G. Wolz, *Chemosphere*, 2015, **129**, 46–53.
46. J. L. Kiplinger, T. G. Richmond and C. E. Osterberg, *Chem. Rev.*, 1994, **94**, 373–431.
47. D. M. Lemal, *J. Org. Chem.*, 2004, **69**, 1–11.
48. J. A. Dean, *Lange's Handbook of Chemistry*, McGraw-Hill, New York, NY, 15th edn, 1999.
49. Z. Wang, I. T. Cousins, U. Berger, K. Hungerbühler and M. Scheringer, *Environ. Int.*, 2016, **89–90**, 235–247.
50. J. C. D'eon and S. A. Mabury, *Environ. Health Perspect.*, 2011, **119**, 344–350.
51. T. Frömel and T. P. Knepper, *Reviews of Environmental Contamination and Toxicology, Perfluorinated Alkylated Substances*, Springer, New York, 2010, vol. 208, pp. 161–177.
52. M. P. Krafft and J. G. Riess, *Chemosphere*, 2015, **129**, 4–19.
53. Z. Wang, I. T. Cousins, M. Scheringer, R. C. Buck and K. Hungerbühler, *Environ. Int.*, 2014, **70**, 62–75.
54. R. C. Buck, P. M. Murphy and M. Pabon, in *Handbook of Environmental Chemistry, Polyfluorinated Chemicals and Transformation Products*, ed. T. P. Knepper and F. T. Lange, Springer-VErlag Berlin Heidelberg, New York, 2012, vol. 17, pp. 1–24.
55. R. E. Banks, B. E. Smart and J. C. Tatlow, *Organofluorine chemistry: Principles and commercial applications*, Plenum, New York, 1994.

56. F. Audenaert, H. Lens, D. Rolly and P. Vander Elst, *J. Text. Inst.*, 1999, **90**, 76–94.
57. A. J. O'Lenick, *Silicones for personal care*, Allured Books, Carol Stream, Il, 2008.
58. L. C. Becker, W. F. Bergfeld, D. V. Belsito, R. A. Hill, C. D. Klaassen, D. Liebler, J. G. Marks, R. C. Shank, T. J. Slaga, P. W. Snyder and F. A. Andersen, *Int. J. Toxicol.*, 2013, **32**, 5S–24S.
59. Y. Fujii, K. H. Harada and A. Koizumi, *Chemosphere*, 2013, **93**, 538–544.
60. L. C. Becker, W. F. Bergfeld, D. V. Belsito, R. A. Hill, C. D. Klaassen, D. C. Liebler, J. G. Marks, R. C. Shank, T. J. Slaga, P. W. Snyder and F. A. Andersen, *Int. J. Toxicol.*, 2014, **33**, 65S–115S.
61. Danish Ministry of Environment and Food, *Risk assessment of fluorinated substances in cosmetic products*, The Danish Environmental Protection Agency, 2018.
62. H. D. Whitehead, M. Venier, Y. Wu, E. Eastman, S. Urbanik, M. L. Diamond, A. Shalin, H. Schwartz-Narbonne, T. A. Bruton, A. Blum, Z. Wang, M. Green, M. Tighe, J. T. Wilkinson, S. McGuinness and G. F. Peaslee, *Environ. Sci. Technol. Lett.*, 2021, **8**, 538–544.
63. G. B. Post, P. D. Cohn and K. R. Cooper, *Environ. Res.*, 2012, **116**, 93–117.
64. S. H. Korzeniowski, R. C. Buck, D. M. Kempisty and M. Pabon, in *Perfluoroalkyl Substances in the Environment*, CRC Press, 2018.
65. C. A. Moody and J. A. Field, *Environ. Sci. Technol.*, 2000, **34**, 3864–3870.
66. E. F. Houtz, C. P. Higgins, J. A. Field and D. L. Sedlak, *Environ. Sci. Technol.*, 2013, **47**, 8187–8195.
67. M. M. Schultz, D. F. Barofsky and J. A. Field, *Environ. Sci. Technol.*, 2004, **38**, 1828–1835.
68. Clean Production Action, GreenScreen Certified™ for Firefighting Foam, https://www.greenscreenchemicals.org/certified/fff-standard (accessed 18 July 2022).
69. G. Yuan, H. Peng, C. Huang and J. Hu, *Environ. Sci. Technol.*, 2016, **50**, 942–950.
70. CEH, *Avoiding hidden hazards: A purchaser's guide to safer foodware*, Center for Environmental Health (CEH), 2018.
71. J. Fassler, The bowls at Chipotle and Sweetgreen are supposed to be compostable, They contain cancer-linked 'forever chemicals', https://newfoodeconomy.org/pfas-forever-chemicals-sweetgreen-chipotle-compostable-biodegradable-bowls/ (accessed 7 August 2019).
72. T. H. Begley, K. White, P. Honigfort, M. L. Twaroski, R. Neches and R. A. Walker, *Food Addit. Contam.*, 2005, **22**(10), 1023–1031.
73. L. S. Lee and H. Trim, *Evaluating perfluoroalkyl acids in composts with compostable food serviceware products in their feedstocks, Summary sheet*, January 2018 (revised March 9, 2018), 2018.
74. X. Trier, K. Granby and J. H. Christensen, *Environ. Sci. Pollut. Res.*, 2011, **18**, 1108–1120.

75. A. C. Blaine, C. D. Rich, E. M. Sedlacko, L. S. Hundal, K. Kumar, C. Lau, M. A. Mills, K. M. Harris and C. P. Higgins, *Environ. Sci. Technol.*, 2014, **48**, 7858–7865.
76. M. C. S. Costello and L. S. Lee, *Curr. Pollut. Rep.*, 2020, 1–21.
77. Clean Production Action, A Guide to Purchasing PFAS-Free Food Service Ware, https://www.cleanproduction.org/resources/entry/pfas (accessed 11 December 2018).
78. Washington State Department of Ecology, *Safer Alternatives to PFAS in Food Packaging. Report to the Legislature pursuant to RCW 70A.222.070*, Olympia, Washington, 2021.
79. Washington State Department of Ecology, *Safer Alternatives to PFAS in Food Packaging. Second report to the Legislature pursuant to RCW 70A.222.070*, Olympia, Washington, 2022.
80. D. M. Kempisty and L. Racz, *Forever Chemicals: Environmental, Economic, and Social Equity Concerns with PFAS in the Environment*, CRC Press, 2021.

CHAPTER 2

Why PFASs Are Chemicals of Concern

A. MARUZZO,*[a] S. A. BĂLAN,[a,b] E. BOXER[a] AND T. A. BRUTON[b]

[a] University of California, Berkeley, School of Public Health, Berkeley, CA, 94704, USA; [b] California Department of Toxic Substances Control, Sacramento, CA, 95812, USA
*Email: aaron_maruzzo@berkeley.edu

2.1 Introduction

In 2001, a class action lawsuit was filed on behalf of some 80 000 people across West Virginia and Ohio for damages resulting from perfluorooctanoic acid (PFOA) drinking water contamination from DuPont's Washington Works plant in Parkersburg, West Virginia. PFOA, a member of the chemical class of perfluoroalkyl and polyfluoroalkyl substances (PFASs), was used by DuPont to manufacture polytetrafluoroethylene (PTFE), a polymeric PFAS used in a myriad of consumer products such as non-stick frying pans, automobile components, and semiconductors. As part of the settlement, an independent panel of three expert epidemiologists was formed to carry out a community study to evaluate if there is a "probable link" between PFOA exposure and any human disease.[1] The panel was referred to as the C8 Science Panel and their work was referred to as the C8 Health Project (C8 is a nickname for PFOA since it contains eight carbon atoms).

From 2005 to 2006, the C8 Health Project collected information from about 69 000 residents near the Washington Works plant through interviews, questionnaires, and blood samples.[2] In 2011 and 2012, the C8 Science Panel delivered probable link findings for six human diseases: high cholesterol, ulcerative colitis, thyroid disease, testicular cancer, kidney cancer, and

Green Chemistry Series No. 81
Toward a PFAS-free Future: Safer Alternatives to Forever Chemicals
Edited by Simona A. Bălan, Thomas A. Bruton and Kimberly G. Hazard

Published by the Royal Society of Chemistry, www.rsc.org

pregnancy-induced hypertension.[3] The landmark study was the first to document potential effects of PFOA at community levels. More broadly, the C8 study helped launch a movement to investigate PFASs – what they are, what hazards they pose, and how they can be managed.

Since the early 2000s, the number of scientific studies and policy initiatives related to PFASs have increased exponentially. Our intent in this chapter is not to present an exhaustive review of the literature, but rather to provide a general overview to set a foundation for the subsequent chapters.

2.2 Overview of PFAS Hazards and Exposures

PFASs are a large and diverse class of highly fluorinated organic chemicals used in numerous industrial processes and consumer products. While precise definitions vary (see Chapter 1), the class is generally thought to contain thousands of unique substances.[4] The stable carbon–fluorine bonds that contribute to the industrial utility of PFASs also make them extremely persistent, which has led to their accumulation in the environment, biota, and humans.[5–25] Some well-studied PFASs, like PFOA and its related compound, perfluorooctanesulfonic acid (PFOS), are strongly associated with adverse human health effects and ecological damage.[16,26] Some PFASs are highly mobile, some are bioaccumulative, and some can contribute to global warming.[27–29]

PFOA and PFOS belong to a relatively small subclass of PFASs called perfluoroalkyl acids (PFAAs). PFAAs are perfluorinated substances, meaning that all their carbon atoms (except those associated with functional groups) are fully fluorinated. This makes them extremely resistant to degradation. Longitudinal studies revealed that the presence of PFOA, PFOS, and other PFAAs in human blood serum may cause increased cholesterol levels; liver toxicity; reproductive and developmental toxicity; cancer of the kidney, testicles, and breast; immunosuppression; and pregnancy-induced hypertension.[26,30] Findings from animal studies showed similar and additional cancer and non-cancer effects on the endocrine, hepatic, reproductive, developmental, and immune systems.[26] Given the known hazards of PFOA, PFOS, and their derivatives, manufacturers replaced them with other PFAAs and PFAA derivatives with fewer than eight perfluorinated carbons (*i.e.*, so-called short-chain PFAAs) or with ether linkages in the molecule (*e.g.*, GenX and ADONA). These replacement PFASs were presumed to be safer because they are less bioaccumulative and are eliminated more readily from the human body. However, these assumptions have been challenged by recent evidence suggesting similar toxic potencies and fates between long- and short-chain PFAAs and their precursors.[31–37]

Fewer studies have examined the potential hazard traits and public health impacts of PFASs other than PFAAs.[38,39] However, the hazards associated with PFAAs are relevant to PFASs as a class, because PFAAs are the terminal degradation products, processing aids, feedstocks, or impurities of nearly all other PFASs.[40] Thus, while not all PFASs are equally bioaccumulative or toxic, most PFASs are highly persistent or degrade into other highly persistent PFASs. This high environmental persistence, along with ubiquitous exposures and potential public health and environmental impacts, has prompted many regulatory agencies to recognize the entire class of PFASs, or a subset of its members, as chemicals of significant concern.

The following sections provide more detail about the hazard traits of PFASs, using the GreenScreen chemical hazard assessment framework as an outline. GreenScreen combines hazard data from authoritative lists, peer-reviewed literature, and estimation techniques to rate chemicals across a range of human health endpoints, environmental hazards, and physical hazards. Standardized criteria are used to categorize a chemical as having very high, high, moderate, low, or unknown levels of concern for each endpoint.[41] Figure 2.1 presents a Green Screen hazard table for five commercially relevant PFASs compiled from publicly available data on the ToxServices website.[42–46]

The GreenScreen group I and II/II* endpoints cover important public health impacts. Group I endpoints (carcinogenicity, mutagenicity, reproductive toxicity, developmental toxicity, and endocrine toxicity) represent chronic, adverse impacts that are potentially induced at low doses, can be transferred between generations, and are commonly used as the basis for regulation. Group II/II* endpoints (acute mammalian toxicity, systemic toxicity, neurotoxicity, sensitization, and irritation) are also necessary for chemical hazard assessment, often appear in specific populations, and can be mitigated.

2.2.1 Human Health Hazards

2.2.1.1 Carcinogenicity

The International Agency for Research on Cancer (IARC) classified tetrafluoroethylene as a probable human carcinogen (group 2A) in 1999, and PFOA as a possible carcinogen (group 2B) in 2016.[26] Supporting evidence for these determinations came from animal studies and studies of highly exposed human populations showing increased risks for kidney, testicular, and prostate cancer.[26] PFOS is suspected to be a carcinogen. Officials at the U.S. Environmental Protection Agency (EPA), as well as in California and Minnesota, identified increased incidences of liver adenomas, pancreatic acinar cell tumors, and testicular Leydig cell adenomas from rat studies.[47–49] However, the lack of population-based data supporting these experimental findings, equivocal rodent-to-human generalizability, and undefined mechanisms of action have constrained other authoritative agencies from classifying PFOS as a carcinogen.

Of the PFASs shown in Figure 2.1, carcinogenicity is rated as moderate or high for hexafluoropropylene, 8:2 fluorotelomer acrylate, and 6:2 FTOH.[42,44,45] For hexafluoropropylene, carcinogenicity scores are based on experimental data for the surrogate compound tetrafluoroethylene.[44] Structural alerts from

	Group I					Group II and II*									Ecotox		Fate		Physical	
Chemical name (CAS No.)	C	M	R	D	E	AT	ST	ST*	N	N*	SnS*	SnR*	IrS	IrE	AA	CA	P	B	Rx	F
Hexafluoropropylene (116-15-4)	*H*	L	DG	DG	DG	**M**	**vH**	*H*	*vH*	*M*	*L*	*L*	*H*	*H*	*M*	*M*	*H*	**vL**	*L*	**L**
3-Ethoxyperfluoro(2-methylhexane) (297730-93-9)	*L*	**L**	*L*	*L*	DG	**L**	**L**	**L**	*L*	*L*	**L**	*L*	**L**	**L**	*L*	*vH*	*vH*	**vH**	*L*	**L**
Perfluorodecyl acrylate, i.e., 8:2 fluorotelomer acrylate (27905-45-9)	*M*	**L**	*L*	*M*	DG	**L**	*M*	*H*	DG	*H*	**L**	*M*	**L**	**L**	*L*	*L*	**vH**	*H*	*L*	**L**
Perfluorohexanoic acid (307-24-4)	**L**	*L*	*L*	*M*	DG	**M**	**H**	*M*	DG	**L**	*L*	DG	*vH*	*vH*	**M**	**L**	**vH**	**L**	*M*	*M*
2-Perfluorohexylethanol, i.e., 6:2 FTOH (647-42-7)	*M*	*L*	**L**	*M*	*M*	**H**	**vH**	*H*	DG	DG	**L**	*L*	**L**	**L**	**H**	*H*	*vH*	**vL**	*L*	**M**

Very Low	Low	Moderate	High	Very High

Figure 2.1 GreenScreen Hazard Assessment for five commonly used PFASs. Data from ref. 42–46 (ToxServices, ToxFMD Screened Chemistry™ Library – FMD, https://database.toxservices.com/FMD/User/Greenscreen/Greenscreen?id=1313). C = carcinogenicity; M = mutagenicity; R = reproductive toxicity; D = developmental toxicity; E = endocrine toxicity; AT = acute (mammalian) toxicity; ST = systemic toxicity; N = neurotoxicity; SnS = skin sensitization; SnR = respiratory sensitization; IrS = skin irritation; IrE = eye irritation; AA = acute aquatic toxicity; CA = chronic aquatic toxicity; P = persistence; B = bioaccumulation potential; Rx = reactivity; F = flammability; BM-1 = Benchmark 1: Avoid – Chemical of High Concern. * Indicates repeated exposure. Italics represents low confidence; and bold represents high confidence.

predictive models identified 8:2 fluorotelomer acrylate and 6:2 FTOH as potential non-genotoxic carcinogens.[42,45]

2.2.1.2 *Mutagenicity*

There is mixed evidence for mutagenicity of PFASs. *In vivo* and *in vitro* toxicity studies for PFOA, perfluorohexanoic acid (PFHxA), perfluorononanoic acid (PFNA), perfluorodecanoic acid (PFDA), PFOS, perfluorohexanesulfonic acid (PFHxS), and perfluorobutanesulfonic acid (PFBS) found no evidence of genotoxicity in bacterial, mammalian, or human hepatoma HepG2 cells.[26,50] However, prior reviews suggest that PFASs may indirectly lead to collateral DNA damage caused by oxidative stress.[50] Other evidence points to possible PFOS-induced DNA damage observed in fish, invertebrates, and certain cell lines.[47] Human studies found correlations between serum PFOS and urinary 8-OhdG, a biomarker for oxidative DNA damage.[47]

2.2.1.3 *Reproductive and Developmental Toxicity*

PFASs may have deleterious effects on pregnancy, birth outcomes, and infant development. There is mixed evidence for an association between PFAA exposure and birth outcomes, especially low birth weight, although systematic reviews have concluded that *in utero* exposure to PFOA and PFOS may cause reduced fetal growth in animals and humans.[26,50] Other PFASs are also present in authoritative lists that identify carcinogens, mutagens, or reproductive or developmental toxicants; these lists include California Proposition 65, the European Union's Registration, Evaluation, Authorization and Restriction of Chemicals (REACH) Annex VII, and the German list MAK.

Of the PFASs shown in Figure 2.1, 8:2 fluorotelomer acrylate, PFHxA, and 6:2 FTOH received a moderate score for developmental toxicity.[42,43,45] Scores were based on observed toxic effects in rats and mice studies, including increased pup mortality, altered pup body weight, and reduced viability.

2.2.1.4 *Endocrine Toxicity*

Many PFASs are suspected to be endocrine disruptors, primarily due to findings which show that PFAS exposure alters thyroid hormone levels. PFASs may reduce circulating total thyroxine levels by competitively binding to thyroid hormone binding proteins.[51] Other *in vitro* evidence suggests that certain PFAAs can interact with estrogen and androgen receptors, perturb cell signaling *via* the hypothalamic–pituitary–adrenal (HPA) axis, and cause effects mediated by estrogen receptor (ER) signaling.[26,50]

Of the five PFASs shown in Figure 2.1, four were not ranked for endocrine toxicity because of data gaps. 6:2 FTOH received moderate score with uncertainty.[45] 6:2 FTOH is present on The Endocrine Disruption Exchange (TEDX) screening list, with uncertainty about whether cancer, reproductive, developmental, or systemic effects are endocrine-mediated.

2.2.1.5 Group II/II* Endpoints

Among group II/II* endpoints, acute and systemic toxicity stand out as of greatest concern (Figure 2.1). Mammalian toxicity assays suggest that acute, short-term, and subchronic PFAS exposure can cause several toxic effects to animals and humans. PFASs tend to accumulate and reach toxic thresholds in protein-rich organs such as the blood, liver, kidney, and thyroid, partly due to preferential binding to lipoproteins or fatty-acid binding proteins.[26,50] Dose-dependent relationships have also been documented for altered gene expression, altered protein levels, cytotoxicity, and histopathological differences, ranging from the cellular scale up to whole organisms.[2]

Of note, immunotoxicity is included under systemic toxicity and may be a sensitive endpoint of interest. Impaired immune function has been observed in both rodents, nonhuman primates, and longitudinal studies of humans exposed to PFAAs.[26,50] For example, prenatal exposure to PFOS and PFHxS may suppress adaptive and immune responses, weakening vaccine effectiveness, and increasing infection risks to diphtheria, rubella, mumps, *Hemophilus*, or respiratory infections.[26,50]

2.2.2 Ecotoxicity

Exposure to PFASs may also cause deleterious impacts on plants, aquatic organisms, and terrestrial wildlife. For the chemicals evaluated in Figure 2.1, acute aquatic toxicity ranged from low to high and chronic aquatic toxicity ranged from low to very high. For a detailed discussion of changes in growth, reproduction, development, and other types of toxicity associated with exposure to different PFASs, see the critical review by Ankley *et al.*[16]

2.2.3 Environmental Persistence

Due to their strong carbon–fluorine bonds, many PFASs have high chemical and thermal stability and do not degrade readily in the environment. Almost all PFASs are either very persistent or degrade into very persistent compounds. In the Green Screen framework, persistence is scored high when a compound has a half-life >40 days in water or >60 days in soil or sediment. The half-lives of some non-polymeric PFASs are estimated to be >40 years in water, and those of some fluoropolymers are estimated to be >1000 years in soil.[52] All five PFASs shown in Figure 2.1 are highly or very highly persistent based on experimental data from biodegradation studies, predicted half-lives, or degradation products known to be highly persistent.[42–46]

Because of their persistence, PFASs released during PFAS manufacturing, or through the manufacturing, use, and disposal of PFAS-containing products, have become ubiquitous environmental contaminants. PFASs are routinely detected in surface water, ground water, air, soil, and sediment samples. Some PFASs are susceptible to long-range atmospheric or oceanic transport that disperses them to remote areas far from where they were produced or used.[53]

Persistence is a significant hazard trait recognized by many regulatory authorities, usually in conjunction with bioaccumulation, toxicity, or both.[54] Several authors have argued that persistence alone should be sufficient as the basis for regulation. This "P-sufficient" approach claims that the continuous manufacture, use, and emissions of persistent pollutants will lead to "widespread, long-lasting, and increasing contamination, which will inevitably result in increasing probabilities of known and unknown effects on human health and the environment.[55]

2.2.4 Bioaccumulation Potential

Traditional, lipid-based predictive methods used to estimate bioaccumulation factors for neutral organic compounds like polychlorinated biphenyls are generally not applicable to PFAAs and other ionizing PFASs due to their tendency to bind to proteins instead of lipids.[16,26,50] Certain PFAAs are known to bioaccumulate significantly in humans, plants, and terrestrial and aquatic organisms. Higher bioaccumulation potentials have been observed among PFAAs with longer chain lengths, and among perfluoroalkanesulfonic acids (PFSAs) such as PFOS compared to perfluoroalkyl carboxylic acids (PFCAs) of the same chain length.[26] Interspecies differences in bioaccumulation exist, with some authors suspecting that bioaccumulation or bioconcentration factors derived from fish are less relevant to non-water-breathing animals.[16,26,50] Field-based measurements among terrestrial mammals indicate that PFCAs have strong biomagnification potential, meaning that PFCAs can accumulate in higher concentrations at higher trophic levels.[56] Plants can also uptake PFASs readily from soil.[57,58]

The reported human serum elimination half-lives for long-chain PFAAs such as PFOA and PFOS are on the order of years.[2] While shorter-chain PFAAs may be eliminated relatively rapidly from the human body, recent studies have found that exposure to short-chain precursors such as 6:2 FTOH may lead to intermediate metabolites that are more persistent in the human body and more toxic than the terminal short-chain PFAA degradation products.[59–61]

Bioaccumulation ranged from very low to very high among the PFASs as shown in Figure 2.1, reflecting a wide range of variability for these endpoints. For instance, measured data obtained from aquatic organisms exposed to 3-ethoxyperfluoro(2-methylhexane) indicate BCF values >5000, which meets the criteria for high bioaccumulation potential.[46] 8:2 fluorotelomer acrylate was ranked high for bioaccumulation because it can degrade into PFOA, which is known to be bioaccumulative.[42,52]

2.2.5 Human Exposure

Nearly all humans are exposed to PFASs. National, regional, and state biomonitoring programs have repeatedly reported widespread detection of PFAAs in human serum samples.[1,62–66] PFOA, PFOS, PFHxS, PFNA, and PFDA were detected in the U.S. general population in >90% of samples between 1999–2000 and 2003–2004.[62,67] Moreover, biomonitoring studies to date have mostly been limited to certain PFAAs, which suggests that the extent of human exposure to PFASs is likely underestimated. Methods targeting extractable or adsorbable organofluorine in blood have recently been explored for the measurement of total organofluorine content.[68] This method enables researchers to detect the presence of unidentified PFASs that may otherwise be unaccounted for, but does not provide any structural information about the chemicals detected.

Humans are exposed to PFAAs through ingestion of contaminated food, water, and dust, inhalation of air, hand-to-mouth behaviors, dermal contact, and transplacental and lactational transfer.[69–72] Food and water consumption, followed by ingestion of house dust, are thought to be the primary routes of exposure.

Certain groups are highly exposed to PFASs compared to background levels in the general population. Sensitivity to the effects of exposure may also vary between different groups. For instance, toddlers, infants, and children ingest more PFASs per body weight than adults from diet, water, and indoor exposures. Other mechanisms of early-life exposure include transplacental transfer, lactational transfer *via* breastmilk, and frequent hand-to-mouth behaviors. These exposures are of concern because early in life, critical developmental periods are more susceptible to disruption and can cause long-term effects later in life. Immune system dysregulation, for instance, has been highly studied among infants and children.[24,73–76] While there are mixed results on whether there is an association between PFAS exposures and allergic reactions and asthma to name a few, there is an indication of immunosuppression, with diminished antibody response to vaccination in children exposed to certain PFASs.[77]

Occupationally exposed populations have higher serum PFAS levels than the general population. These groups include workers involved in the manufacture, processing, or use of PFASs or PFAS-containing materials, including fluorochemical manufacturing workers, firefighters, and applicators of PFAS-containing products (*e.g.*, ski waxers).[78–81] Elevated serum PFASs have been reported among seafood consuming populations, including Inuit men in Greenland, whaling men in the Faroe Islands, and commercial fishery employees in China.[24] Residents in communities near major facilities that use or manufacture PFASs also tend to have highly elevated PFAS levels.[82–85]

2.3 Consumer Product Regulations

Growing awareness of PFAS hazards and exposures have prompted some governments to restrict the use of PFASs in consumer products. Consumer product regulations serve as a major leverage point to alter chemical manufacture, use, processing, release, and emissions, and can have significant impact on consumer exposure to chemicals of concern. For instance, follow-up national biomonitoring surveys found a decrease in the serum levels of PFOS and PFOA from 1999–2000 to 2015 by 84 and 70%, respectively, coincident with the phase out of these substances during the early 2000s. However, the widespread use of diverse PFASs across product sectors, scientific data gaps, and the complex decision-making frameworks that underlie chemical regulation have resulted in a complex patchwork of regulatory initiatives in the U.S. and abroad.

Most chemical regulations are risk-based, assessing and managing the risks of exposures on a chemical-by-chemical basis. Two emerging frameworks aim to apply more holistic and precautionary approaches, incentivizing protective efforts despite incomplete information. Modeled after the term's use in the Montreal Protocol, the essential-use approach proposes to restrict or ban the non-essential uses of chemicals of concern such as PFASs.[55,86,87] According to Cousins *et al.* (2019), examples of non-essential uses of PFASs include dental floss, guitar strings, cosmetics, food packaging, carpets, and firefighting foam.[88] In addition to the essential use approach, many authors have argued that the entire PFAS class, or broad subclasses of PFASs, should be targeted for regulation instead of only individual compounds.[39,89,90] This class-based approach shifts the primary concern from risks toward hazards. A growing global community of scientists, community-based organizations, environmental groups, regulators, and political leaders favors the class-based approach,[54,91–93] with some resistance from the fluorochemical industry.[94]

2.3.1 U.S. Federal Regulations

U.S. federal agencies with regulatory authority over toxic chemicals in consumer products include the Consumer Product Safety Commission, the Occupational Health and Safety Administration, and the Environmental Protection Agency (EPA). To date, EPA has been the most active agency in addressing PFASs. EPA's primary authority for regulating consumer product chemicals comes from the Toxic Substances Control Act (TSCA).

Historically, EPA's approach to regulating the production and use of PFASs has relied on a combination of voluntary agreements with manufacturers and regulations restricting future uses of those PFASs that have already been phased out.[95] Most of this work has focused specifically on long-chain PFAAs and their precursors. After the phaseout of PFOS and related long-chain chemistries by the 3M Company between 2000 and 2002, EPA published two significant new use rules (SNURs) requiring anyone intending to manufacture or import these chemicals to notify EPA beforehand.[95] These two 2002 SNURs applied collectively to 88 individual PFASs. Starting in 2006, EPA worked with eight major fluorochemical manufacturers to voluntarily phase out the production of PFOA and related chemistries. The 2010/2015 PFOA Stewardship Program aimed to completely eliminate PFOA, PFOA precursors, and higher homologues from facility emissions and products by 2015.[95] In tandem, EPA proposed a SNUR in 2015 requiring notification prior to manufacture or import of PFOA and related chemistries. Other SNURs proposed or finalized by EPA address the use of PFOA and related chemistries in carpet, the import of long-chain PFASs as surface coatings on articles, and new uses of long-chain PFASs that have previously been phased out in the United States.[95] EPA is also working to address PFASs under other U.S. environmental laws that regulate hazardous waste, water quality, air quality, *etc.* While these laws largely do not pertain to the use of chemicals in consumer products, EPA's action in these areas may increase the real or perceived liability associated with the consumer product uses of PFASs.

Several bills aiming to restrict the use of PFASs in certain products have been proposed in the U.S. Congress, including legislation to ban the use of PFASs in food packaging and cosmetics.[96,97] Most of these proposals have not been enacted. One notable provision of the 2020 National Defense Authorization Act banned the use of PFASs in packaging for Meals Ready to Eat (MREs) used by the military.[98]

2.3.2 U.S. State Actions

Numerous U.S. states have taken their own legislative or regulatory actions pertaining to PFASs.[99–101] Over 30 states have adopted or are currently proposing laws to address PFASs in consumer products and the environment.[101] As of early 2023, a total of 104 policies have been enacted targeting PFASs.[102] Of these, over 25 policies aim to phase out or restrict the use of PFASs in certain products.[102]

We highlight efforts in California, Washington, and Maine as examples of diverse state actions on consumer products containing PFASs. Of note, certain state actions successfully took a class-based approach to PFAS regulation in consumer products. In California, the legislature banned all PFASs in plant-based food packaging (AB 1200),[103] children's products (AB 652),[104] textiles (AB 1817),[105] cosmetics (AB 2771),[106] and firefighting foam (SB 1044).[107] Additionally, California's Department of Toxic Substances Control listed carpets and rugs containing PFASs as well as treatments containing PFASs for use on converted textiles or leathers as Priority Products under the Safer Consumer Products Regulations. This requires the manufacturers of these products to either evaluate safer alternatives or stop selling the PFAS-containing products in California. Moreover, the California Office of Environmental Health Hazard Assessment added three PFASs (PFOA, PFOS, and PFNA) to its Proposition 65 list of chemicals known to the state of California as carcinogens or toxicants to the reproductive or developmental system, which means that businesses are required to disclose the presence of these PFASs in their products.

Washington State has also proactively addressed PFASs as a class, targeting consumer products suspected of the highest relative sources of exposure: food packaging, after-market leather and textile treatments, carpets and rugs, and leather and textile furnishings.[108] Maine now also requires manufacturers of products containing PFASs to report these to the state's Department of Environmental Protection and prohibits the sale of carpets or rugs, as well as the sale of fabric treatments, that contain intentionally added PFAS. Beginning January 1, 2030, any product containing intentionally added PFASs may not be sold in Maine unless the use of PFASs in the product is specifically designated as a currently unavoidable use by the Maine Department of Environmental Protection.[109]

2.3.3 International Regulations

Several other countries besides the U.S. are actively pursuing efforts to reduce the use of PFASs. Here we highlight a few examples.

In 2009, PFOS was listed under Annex B (restriction) of the Stockholm Convention on Persistent Organic Pollutants,[110] meaning that signatories must pursue efforts to restrict its production and use. In 2019, PFOA and its related compounds were listed on Annex A (elimination), while long-chain PFCAs, PFHxS, and PFHxS-related compounds were recommended for listing in Annex A.[110] Among the 152 countries that have ratified the Stockholm Convention, regulations have been uneven. For example, Denmark banned fluorinated substances in paper and board food contact materials,[111] while India has yet not implemented any restrictions on PFASs in any environmental media or consumer products.[112] Notably, the U.S. has not ratified the Stockholm Convention.

In 2011, China restricted PFOS and PFOA production and incentivized further research into the extent of contamination and on alternatives.[113] In 2018, South Australia became the first jurisdiction to ban PFAS-containing firefighting foam. In 2021, Japan banned PFOA and its salts in the manufacture and importation of various consumer products including paper, textiles, fire extinguishers, and floor waxes.[114]

The European Union (EU) has placed several PFASs on the REACH Regulation's Candidate List of substances of very high concern (SVHC) due to their persistent, bioaccumulative, and toxic (PBT) and very persistent and very bioaccumulative (vPvB) properties. Once a substance is included in the SVHC list, it sends a strong signal to companies using an SVHC that they should start looking for less hazardous alternatives. Recently the EU recognized that persistent substances that are mobile in the environment pose an equivalent level of concern, and a PFAS not considered bioaccumulative was declared an SVHC on the basis of its persistence and mobility.[115] The EU has also agreed on harmonized classifications for several PFASs under the classification, labelling, and packaging (CLP) regulation. In addition, the EU is taking measures to introduce these classifications (PBT; vPvB; persistent, mobile, and toxic (PMT); and very persistent and very mobile (vPvM)) into the Globally Harmonized System of Classification and Labelling of Chemicals (GHS), to get global recognition of persistence as an environmental hazard.[116]

The REACH Regulation also provides a basis for restricting chemicals, and a number of PFASs, including PFOS and PFOA, are now banned in finished and semi-finished products. Several other ambitious efforts aimed at regulating PFASs are underway in the EU. In 2020, the European Commission published the Chemicals Strategy for Sustainability, including strategies to phase out all non-essential uses of PFASs in the EU.[117,118] This led to a proposal prepared by Denmark, Germany, the Netherlands, Norway, and Sweden for a restriction under REACH that would ban all uses of all PFASs unless an exemption has been granted for a specific use.[119] The proposed restriction is expected to undergo a year-long rulemaking process in 2023, which will include scrutiny by the European Chemicals Agency (ECHA) and its committees on risk assessment and on socio-economic assessment. Final adoption is expected sometime in 2024.

2.3.4 Future Directions for PFAS Regulation

Dialogue among the research, social, and political communities regarding PFAS issues is increasing, with growing pressure on regulators and policymakers to take action and on industry to transition away from PFASs. While many regulatory efforts to date have been risk-based and focused on select PFAAs, the regulatory landscape for PFASs is rapidly changing both in the U.S. and internationally. Recent regulatory action has been moving toward managing PFASs as a class, particularly as new science uncovers more about the hazard traits associated with these compounds. Emerging technologies capable of measuring broader classes of PFASs in environmental media may also lead to more hazard-based, holistic, and precautionary approaches to PFASs. Therefore, there is, and will continue to be, a growing need to identify safer alternatives to PFASs. The following chapters provide guidance and examples for doing so.

Abbreviations

6:2 FTOH	6:2 Fluorotelomer alcohol
ADONA	Trade name for 4,8-dioxa-3*H*-perfluorononanoate
CLP	Classification, labelling, and packaging regulation (EU)
ECHA	European Chemicals Agency
EPA	U.S. Environmental Protection Agency
EU	European Union
ER	Estrogen receptor
GenX	Trade name for HFPO-DA (hexafluoropropylene oxide dimer acid)
HPA	Hypothalamic–pituitary–adrenal axis
MREs	Meals ready to eat
PFAA	Perfluoroalkyl acids
PFAS	Perfluoroalkyl and polyfluoroalkyl substances
PFBS	Perfluorobutanesulfonic acid
PFCA	Perfluoroalkyl carboxylic acid
PFDA	Perfluorodecanoic acid
PFHxA	Perfluorohexanoic acid

PFHxS	Perfluorohexanesulphonic acid
PFNA	Perfluorononanoic acid
PFOA	Perfluorooctanoic acid
PFOS	Perfluorooctanesulfonic acid
REACH	European Union – Registration, Evaluation, Authorization and Restriction of Chemicals
SNUR	Significant new use rules
SVHC	Substances of very high concern (REACH candidate list)
TEDX	Thc Endocrinc Disruption Exchangc
TSCA	Toxic Substances Control Act

Acknowledgements

The authors would like to thank Chris Leonetti, Nancy Ostrom, and Gretta Goldenman for their helpful comments on the manuscript.

References

1. S. J. Frisbee, A. P. Brooks, A. Maher, P. Flensborg, S. Arnold, T. Fletcher, K. Steenland, A. Shankar, S. S. Knox, C. Pollard, J. A. Halverson, V. M. Vieira, C. Jin, K. M. Leyden and A. M. Ducatman, *Environ. Health Perspect.*, 2009, **117**, 1873–1882.
2. Background Information on Lawsuit Settlement, http://www.c8sciencepanel.org/panel.html (accessed 25 October 2022).
3. C8 Probable Link Reports, http://www.c8sciencepanel.org/prob_link.html (accessed 25 October 2022).
4. E. Hammel, T. F. Webster, R. Gurney and W. Heiger-Bernays, *iScience*, 2022, **25**, 104020.
5. X. C. Hu, D. Q. Andrews, A. B. Lindstrom, T. A. Bruton, L. A. Schaider, P. Grandjean, R. Lohmann, C. C. Carignan, A. Blum, S. A. Balan, C. P. Higgins and E. M. Sunderland, *Environ. Sci. Technol. Lett.*, 2016, **3**, 344–350.
6. G. R. Johnson, M. L. Brusseau, K. C. Carroll, G. R. Tick and C. M. Duncan, *Sci. Total Environ.*, 2022, **841**, 156602.
7. A. L. Jarvis, J. R. Justice, M. C. Elias, B. Schnitker and K. Gallagher, *Environ. Toxicol. Chem.*, 2021, **40**, 2425–2442.
8. B. Xu, S. Liu, J. L. Zhou, C. Zheng, J. Weifeng, B. Chen, T. Zhang and W. Qiu, *J. Hazard. Mater.*, 2021, **412**, 125159.
9. L. Ahrens, *J. Environ. Monit.*, 2011, **13**, 20–31.
10. F. Xiao, *Water Res.*, 2017, **124**, 482–495.
11. M. Strynar, S. Dagnino, R. McMahen, S. Liang, A. Lindstrom, E. Andersen, L. McMillan, M. Thurman, I. Ferrer and C. Ball, *Environ. Sci. Technol.*, 2015, **49**, 11622–11630.
12. M. J. Strynar, A. B. Lindstrom, S. F. Nakayama, P. P. Egeghy and L. J. Helfant, *Chemosphere*, 2012, **86**, 252–257.

13. K. Rankin, S. A. Mabury, T. M. Jenkins and J. W. Washington, *Chemosphere*, 2016, **161**, 333–341.
14. X. Bai and Y. Son, *Sci. Total Environ.*, 2021, **751**, 141622.
15. Q. Wang, M. M. P. Tsui, Y. Ruan, H. Lin, Z. Zhao, J. P. H. Ku, H. Sun and P. K. S. Lam, *Chemosphere*, 2019, **231**, 468–477.
16. G. T. Ankley, P. Cureton, R. A. Hoke, M. Houde, A. Kumar, J. Kurias, R. Lanno, C. McCarthy, J. Newsted, C. J. Salice, B. E. Sample, M. S. Sepúlveda, J. Steevens and S. Valsecchi, *Environ. Toxicol. Chem.*, 2021, **40**, 564–605.
17. J. P. Giesy and K. Kannan, *Environ. Sci. Technol.*, 2002, **36**, 146A–152A.
18. J. P. Giesy and K. Kannan, *Environ. Sci. Technol.*, 2001, **35**, 1339–1342.
19. L. L. Stahl, B. D. Snyder, A. R. Olsen, T. M. Kincaid, J. B. Wathen and H. B. McCarty, *Sci. Total Environ.*, 2014, **499**, 185–195.
20. B. C. Kelly, M. G. Ikonomou, J. D. Blair, B. Surridge, D. Hoover, R. Grace and F. A. P. C. Gobas, *Environ. Sci. Technol.*, 2009, **43**, 4037–4043.
21. G. Boisvert, C. Sonne, F. F. Rigét, R. Dietz and R. J. Letcher, *Environ. Pollut.*, 2019, **252**, 1335–1343.
22. R. Ghisi, T. Vamerali and S. Manzetti, *Environ. Res.*, 2019, **169**, 326–341.
23. S. H. Brandsma, J. C. Koekkoek, M. J. M. van Velzen and J. de Boer, *Chemosphere*, 2019, **220**, 493–500.
24. E. M. Sunderland, X. C. Hu, C. Dassuncao, A. K. Tokranov, C. C. Wagner and J. G. Allen, *J. Exposure Sci. Environ. Epidemiol.*, 2019, **29**, 131–147.
25. H. Fromme, S. A. Tittlemier, W. Völkel, M. Wilhelm and D. Twardella, *Int. J. Hyg. Environ. Health*, 2009, **212**, 239–270.
26. ATSDR (Agency for Toxic Substances and Disease Registry), *Toxicological profile for Perfluoroalkyls*, U.S. Department of Health and Human Services, Public Health Service, Atlanta, GA, 2021.
27. K. Prevedouros, I. T. Cousins, R. C. Buck and S. H. Korzeniowski, *Environ. Sci. Technol.*, 2006, **40**, 32–44.
28. C. A. Ng and K. Hungerbühler, *Environ. Sci. Technol.*, 2014, **48**, 4637–4648.
29. W.-T. Tsai, *Environ. Int.*, 2009, **35**, 418–424.
30. E. National Academies of Sciences, *Guidance on PFAS Exposure, Testing, and Clinical Follow-Up*, 2022.
31. Z. Wang, I. T. Cousins, M. Scheringer and K. Hungerbuehler, *Environ. Int.*, 2015, **75**, 172–179.
32. J. M. Conley, C. S. Lambright, N. Evans, M. J. Strynar, J. McCord, B. S. McIntyre, G. S. Travlos, M. C. Cardon, E. Medlock-Kakaley, P. C. Hartig, V. S. Wilson and L. E. Gray, *Environ. Health Perspect.*, 2019, **127**, 37008.
33. M. I. Gomis, R. Vestergren, D. Borg and I. T. Cousins, *Environ. Int.*, 2018, **113**, 1–9.
34. C. M. Butt, D. C. G. Muir and S. A. Mabury, *Environ. Toxicol. Chem.*, 2014, **33**, 243–267.
35. G. Shi, Q. Cui, Y. Pan, N. Sheng, Y. Guo and J. Dai, *Aquat. Toxicol.*, 2017, **190**, 53–61.
36. A. A. Rand and S. A. Mabury, *Toxicology*, 2017, **375**, 28–36.

37. P. Mukerji, J. C. Rae, R. C. Buck and J. C. O'Connor, *Toxicol. Rep.*, 2015, **2**, 130–143.
38. C. A. McDonough, J. L. Guelfo and C. P. Higgins, *Curr. Opin. Environ. Sci. Health*, 2019, **7**, 13–18.
39. Z. Wang, J. C. DeWitt, C. P. Higgins and I. T. Cousins, *Environ. Sci. Technol.*, 2017, **51**, 2508–2518.
40. OECD (Organisation for Economic Co-operation and Development), *Fact Cards of Major Groups of Per- and Polyfluoroalkyl Substances (PFASs)*, 2022.
41. Clean Production Action, GreenScreen® for Safer Chemicals, https://www.greenscreenchemicals.org/static/ee_images/uploads/resources/2pager_greenscreen_2014.pdf (accessed 9 January 2023).
42. ToxServices LLC, *Perfluorodecyl acrylate (CAS #27905-45-9) GreenScreen for Safer Chemicals (GreenScreen) Assessment*, 2016.
43. ToxServices LLC, *Perfluorohexanoic acid (CAS #307-24-4) GreenScreen for Safer Chemicals (GreenScreen) Assessment*, 2016.
44. ToxServices LLC, *Hexafluoropropylene (CAS #116-15-4) GreenScreen for Safer Chemicals (GreenScreen) Assessment*, 2018.
45. ToxServices LLC, *2-Perfluorohexylethanol (CAS #647-42-7) GreenScreen for Safer Chemicals (GreenScreen) Assessment*, 2019.
46. ToxServices LLC, *3-Ethoxyperfluoro(2-methylhexane) (CAS #297730-93-9) GreenScreen for Safer Chemicals (GreenScreen) Assessment*, 2020.
47. OEHHA, *Proposition 65: Evidence on the Carcinogenicity of Perfluorooctane Sulfonate (PFOS) and Its Salts and Transformation and Degradation Products*, California EPA, 2021.
48. U.S. Environmental Protection Agency, Drinking Water Health Advisory for Perfluorooctane Sulfonate (PFOS), 2016, https://www.epa.gov/sites/default/files/2016-05/documents/pfos_health_advisory_final-plain.pdf.
49. Minnesota Department of Health, Toxicological Summary for: Perfluorooctanoate, 2020, https://www.health.state.mn.us/communities/environment/risk/docs/guidance/gw/pfoa.pdf#:~:text=In%202017%20MDH%20completed%20a%20focused%20re-evaluation%20that,The%20basis%20of%20the%20value%20is%20presented%20below.
50. S. E. Fenton, A. Ducatman, A. Boobis, J. C. DeWitt, C. Lau, C. Ng, J. S. Smith and S. M. Roberts, *Environ. Toxicol. Chem.*, 2021, **40**, 606–630.
51. F. Coperchini, L. Croce, G. Ricci, F. Magri, M. Rotondi, M. Imbriani and L. Chiovato, *Front. Endocrinol.*, 2021, **11**, 612320.
52. J. W. Washington, J. J. Ellington, T. M. Jenkins, J. J. Evans, H. Yoo and S. C. Hafner, *Environ. Sci. Technol.*, 2009, **43**, 6617–6623.
53. S. Kurwadkar, J. Dane, S. R. Kanel, M. N. Nadagouda, R. W. Cawdrey, B. Ambade, G. C. Struckhoff and R. Wilkin, *Sci. Total Environ.*, 2022, **809**, 151003.
54. S. A. Bălan, V. C. Mathrani, D. F. Guo and A. M. Algazi, *Environ. Health Perspect.*, 2021, **129**, 025001.
55. I. T. Cousins, C. A. Ng, Z. Wang and M. Scheringer, *Environ. Sci.: Processes Impacts*, 2019, **21**, 781–792.

56. R. Grønnestad, B. P. Vázquez, A. Arukwe, V. L. B. Jaspers, B. M. Jenssen, M. Karimi, J. L. Lyche and Å. Krøkje, *Environ. Sci. Technol.*, 2019, **53**, 13390–13397.
57. M. C. S. Costello and L. S. Lee, *Curr. Pollut. Rep.*, 2020, 1–21.
58. W. Wang, G. Rhodes, J. Ge, X. Yu and H. Li, *Chemosphere*, 2020, **261**, 127584.
59. R. C. Buck, J. Franklin, U. Berger, J. M. Conder, I. T. Cousins, P. de Voogt, A. A. Jensen, K. Kannan, S. A. Mabury and S. P. van Leeuwen, *Integr. Environ. Assess. Manage.*, 2011, **7**, 513–541.
60. S. V. Kabadi, J. Fisher, J. Aungst and P. Rice, *Food Chem. Toxicol.*, 2018, **112**, 375–382.
61. P. A. Rice, J. Aungst, J. Cooper, O. Bandele and S. V. Kabadi, *Food Chem. Toxicol.*, 2020, **138**, 111210.
62. A. M. Calafat, Z. Kuklenyik, J. A. Reidy, S. P. Caudill, J. S. Tully and L. L. Needham, *Environ. Sci. Technol.*, 2007, **41**, 2237–2242.
63. A. M. Calafat, L.-Y. Wong, Z. Kuklenyik, J. A. Reidy and L. L. Needham, *Environ. Health Perspect.*, 2007, **115**, 1596–1602.
64. Minnesota Department of Health, *East Metro PFC Biomonitoring Follow-up Project: December 2011 Report to the Community*, 2011.
65. C. H. Yu, C. D. Riker, S. Lu and Z. (Tina) Fan, *Int. J. Hyg. Environ. Health*, 2020, **223**, 34–44.
66. N. M. DeLuca, K. Thomas, A. Mullikin, R. Slover, L. W. Stanek, A. N. Pilant and E. A. Cohen Hubal, *J. Exposure Sci. Environ. Epidemiol.*, 2023, 1–15.
67. A. M. Calafat, L.-Y. Wong, Z. Kuklenyik, J. A. Reidy and L. L. Needham, *Environ. Health Perspect.*, 2007, **115**, 1596–1602.
68. R. Aro, U. Eriksson, A. Kärrman, K. Jakobsson and L. W. Y. Yeung, *Environ. Int.*, 2022, **159**, 107035.
69. D. Trudel, L. Horowitz, M. Wormuth, M. Scheringer, I. T. Cousins and K. Hungerbühler, *Risk Anal.*, 2008, **28**, 251–269.
70. E. M. Sunderland, X. C. Hu, C. Dassuncao, A. K. Tokranov, C. C. Wagner and J. G. Allen, *J. Exposure Sci. Environ. Epidemiol.*, 2019, **29**, 131–147.
71. M. E. Morales-McDevitt, J. Becanova, A. Blum, T. A. Bruton, S. Vojta, M. Woodward and R. Lohmann, *Environ. Sci. Technol. Lett.*, 2021, **8**, 897–902.
72. X. C. Hu, C. Dassuncao, X. Zhang, P. Grandjean, P. Weihe, G. M. Webster, F. Nielsen and E. M. Sunderland, *Environ. Health*, 2018, **17**, 11.
73. K. Winkens, R. Vestergren, U. Berger and I. T. Cousins, *Emerging Contam.*, 2017, **3**, 55–68.
74. J. L. Domingo and M. Nadal, *Environ. Res.*, 2019, **177**, 108648.
75. K. M. Rappazzo, E. Coffman and E. P. Hines, *Int. J. Environ. Res. Public Health*, 2017, **14**, E691.
76. H. M. Goeden, C. W. Greene and J. A. Jacobus, *J. Exposure Sci. Environ. Epidemiol.*, 2019, **29**, 183–195.

77. H. von Holst, P. Nayak, Z. Dembek, S. Buehler, D. Echeverria, D. Fallacara and L. John, *Heliyon*, 2021, **7**, e08160.
78. G. W. Olsen, J. M. Burris, D. J. Ehresman, J. W. Froehlich, A. M. Seacat, J. L. Butenhoff and L. R. Zobel, *Environ. Health Perspect.*, 2007, **115**, 1298–1305.
79. G. Costa, S. Sartori and D. Consonni, *J. Occup. Environ. Med.*, 2009, **51**, 364–372.
80. J. Trowbridge, R. R. Gerona, T. Lin, R. A. Rudel, V. Bessonneau, H. Buren and R. Morello Frosch, *Environ. Sci. Technol.*, 2020, **54**, 3363–3374.
81. B. I. Freberg, L. S. Haug, R. Olsen, H. L. Daae, M. Hersson, C. Thomsen, S. Thorud, G. Becher, P. Molander and D. G. Ellingsen, *Environ. Sci. Technol.*, 2010, **44**, 7723–7728.
82. Y. Li, T. Fletcher, D. Mucs, K. Scott, C. H. Lindh, P. Tallving and K. Jakobsson, *Occup. Environ. Med.*, 2018, **75**, 46–51.
83. R. R. Worley, S. M. Moore, B. C. Tierney, X. Ye, A. M. Calafat, S. Campbell, M. B. Woudneh and J. Fisher, *Environ. Int.*, 2017, **106**, 135–143.
84. J. R. Pritchett, J. L. Rinsky, B. Dittman, A. Christensen, R. Langley, Z. Moore, A. T. Fleischauer, K. Koehler, A. M. Calafat, R. Rogers, L. Esters, R. Jenkins, F. Collins, D. Conner and P. Breysse, *Morb. Mortal. Wkly. Rep.*, 2019, **68**, 647–648.
85. L. A. Darrow, C. R. Stein and K. Steenland, *Environ. Health Perspect.*, 2013, **121**, 1207–1213.
86. J. Glüge, R. London, I. T. Cousins, J. DeWitt, G. Goldenman, D. Herzke, R. Lohmann, M. Miller, C. A. Ng, S. Patton, X. Trier, Z. Wang and M. Scheringer, *Environ. Sci. Technol.*, 2022, **56**, 6232–6242.
87. S. A. Bălan, D. Q. Andrews, A. Blum, M. L. Diamond, S. R. Fernández, E. Harriman, A. B. Lindstrom, A. Reade, L. Richter, R. Sutton, Z. Wang and C. F. Kwiatkowski, *Environ. Sci. Technol.*, 2023, **57**, 1568–1575.
88. I. T. Cousins, G. Goldenman, D. Herzke, R. Lohmann, M. Miller, C. A. Ng, S. Patton, M. Scheringer, X. Trier, L. Vierke, Z. Wang and J. C. DeWitt, *Environ. Sci.: Processes Impacts*, 2019, **21**, 1803–1815.
89. C. F. Kwiatkowski, D. Q. Andrews, L. S. Birnbaum, T. A. Bruton, J. C. DeWitt, D. R. U. Knappe, M. V. Maffini, M. F. Miller, K. E. Pelch, A. Reade, A. Soehl, X. Trier, M. Venier, C. C. Wagner, Z. Wang and A. Blum, *Environ. Sci. Technol. Lett.*, 2020, **7**, 532–543.
90. I. T. Cousins, J. C. DeWitt, J. Glüge, G. Goldenman, D. Herzke, R. Lohmann, M. Miller, C. A. Ng, M. Scheringer, L. Vierke and Z. Wang, *Environ. Sci.: Processes Impacts*, 2020, **22**, 1444–1460.
91. A. Blum, S. A. Balan, M. Scheringer, X. Trier, G. Goldenman, I. T. Cousins, M. Diamond, T. Fletcher, C. Higgins, A. E. Lindeman, P. Graham, P. de Voogt, Z. Wang and R. Weber, *Environ. Health Perspect.*, 2015, **123**, A107–A111.
92. A. Ritscher, Z. Wang, M. Scheringer, J. M. Boucher, L. Ahrens, U. Berger, S. Bintein, S. K. Bopp, D. Borg, A. M. Buser, I. Cousins, J. DeWitt, T. Fletcher, C. Green, D. Herzke, C. Higgins, J. Huang, H. Hung,

T. Knepper, C. S. Lau, E. Leinala, A. B. Lindstrom, J. Liu, M. Miller, K. Ohno, N. Perkola, Y. Shi, L. Småstuen Haug, X. Trier, S. Valsecchi, K. van der Jagt and L. Vierke, *Environ. Health Perspect.*, 2018, **126**, 84502.
93. M. Scheringer, X. Trier, I. T. Cousins, P. de Voogt, T. Fletcher, Z. Wang and T. F. Webster, *Chemosphere*, 2014, **114**, 337–339.
94. J. S. Bowman, *Environ. Health Perspect.*, 2015, **123**, A112–A113.
95. U.S. EPA, EPA Actions to Address PFAS, https://www.epa.gov/pfas/epa-actions-address-pfas (accessed 6 July 2022).
96. M. Hassan, *S.3169 – Keep Food Containers Safe from PFAS Act of 2021*, 2021.
97. S. M. Collins, *No PFAS in Cosmetics Act*, 2021.
98. J. Inhofe, *S.1790 – 116th Congress (2019–2020): National Defense Authorization Act for Fiscal Year 2020*, 2019.
99. J. Kindschuch, T. Lee, E. Voyen and B. C. L. Paisner, PFAS Update, https://www.jdsupra.com/legalnews/pfas-update-current-state-by-state-1938319/ (accessed 12 June 2022).
100. Toxics Use Reduction Institute, Per- and Poly-fluoroalkyl Substances (PFAS): Policy Analysis, 2021, https://www.turi.org/content/download/13639/207519/file/PFAS%20Policy%20Analysis%20May%202021.pdf.
101. Safer States, PFAS, https://www.saferstates.org/toxic-chemicals/pfas/ (accessed 30 June 2022).
102. Safer States, Bill Tracker, https://www.saferstates.org/bill-tracker/ (accessed 12 January 2023).
103. *AB-1200 Plant-based food packaging: cookware: hazardous chemicals.*
104. *AB-652 Product safety: juvenile products: chemicals: perfluoroalkyl and polyfluoroalkyl substances.*
105. *AB-1817 Product safety: textile articles: perfluoroalkyl and polyfluoroalkyl substances (PFAS).*
106. *AB-2771 Cosmetic products: safety.*
107. *SB-1044 Firefighting equipment and foam: PFAS chemicals.*
108. WA State Department of Ecology, Per- and polyfluoroalkyl substances, https://ecology.wa.gov/Waste-Toxics/Reducing-toxic-chemicals/Addressing-priority-toxic-chemicals/PFAS (accessed 12 November 2022).
109. Maine Department of Environmental Protection, PFAS in Products, https://www.maine.gov/dep/spills/topics/pfas/PFAS-products/index.html (accessed 11 January 2023).
110. UNEP (United Nations Environment Programme), All POPs listed in the Stockholm Convention, http://chm.pops.int/TheConvention/ThePOPs/AllPOPs/tabid/2509/Default.aspx (accessed 11 June 2022).
111. Ministry of Environment and Food of Denmark, Danish Veterinary and Food Administration, Ban on fluorinated substances in paper and board food contact materials (FCM), Fact sheet, 2020, https://en.foedevarestyrelsen.dk/Media/638210239823191854/Faktaark%20FCM%20(english).pdf.
112. A. Mukherjee Das and R. Janardhanan, *Environ. Epidemiol.*, 2022, **6**, e199.

113. A. Kenji, China releases draft plan for new chemical pollutants management, https://enviliance.com/regions/east-asia/cn/report_4504 (accessed 1 July 2022).
114. Chemical Watch, Japan to ban PFOA and its salts from October, https://chemicalwatch.com/209342/japan-to-ban-pfoa-its-salts-and-compounds-from-october (accessed 1 July 2022).
115. European Environment Agency, Emerging chemical risks in Europe—'PFAS', https://www.eea.europa.eu/publications/emerging-chemical-risks-in-europe (accessed 17 February 2023).
116. European Commission, Proposal for a Regulation of the European Parliament and of the Council amending Regulation (EC) No. 1272/2008 of the European Parliament and of the Council on classification, labelling and packaging of substances and mixtures, https://ec.europa.eu/info/law/better-regulation/have-your-say/initiatives/12975-Revision-of-EU-legislation-on-hazard-classification-labelling-and-packaging-of-chemicals_en (accessed 17 February 2023).
117. European Commission, Commission adopts new Chemicals Strategy, https://ec.europa.eu/commission/presscorner/detail/en/ip_20_1839 (accessed 7 July 2022).
118. European Commission, Sustainable Chemicals, https://ec.europa.eu/growth/news/sustainable-chemicals-commission-advances-work-restrictions-harmful-chemical-substances-2022-04-25_en (accessed 7 July 2022).
119. ECHA, ECHA receives PFASs restriction proposal from five national authorities, https://echa.europa.eu/-/echa-receives-pfass-restriction-proposal-from-five-national-authorities (accessed 8 February 2023).

CHAPTER 3

Alternatives to PFASs in Molded Fiber Fast Food Packaging

M. TELI,[a] A. MARUZZO[b] AND S. A. BĂLAN*[b,c]

[a] Department of Civil and Environmental Engineering, University of California, Berkeley, CA, 94720, USA; [b] University of California, Berkeley, School of Public Health, Berkeley, CA, 94704, USA; [c] California Department of Toxic Substances Control, Sacramento, CA, 95812, USA
*Email: simona.balan@dtsc.ca.gov

3.1 Challenge Overview

Manufacturers of fast food packaging are faced with a conundrum: environmentally conscious consumers are demanding less single-use petrochemical-based plastics, but most composting facilities do not readily accept bioplastics marketed as compostable.[1] Plant fiber-based food packaging, such as paper, paperboard, and molded fibers, offers a solution. It has the potential to be more sustainably produced, degrades faster than its bioplastic counterparts, and has a natural look and feel that appeals to consumers. However, plant fiber materials require additives to achieve water and oil resistance. Perfluoroalkyl and polyfluoroalkyl substances (PFASs) dominate the industry as inexpensive and versatile additives that can impart water and oil repellency to plant fiber-based food contact materials. However, this convenience comes at a high cost to human health and the environment.[2]

While fast food packaging has a useful life on the order of hours, PFASs have earned the nickname "forever chemicals" because some can persist

Green Chemistry Series No. 81
Toward a PFAS-free Future: Safer Alternatives to Forever Chemicals
Edited by Simona A. Bălan, Thomas A. Bruton and Kimberly G. Hazard

Published by the Royal Society of Chemistry, www.rsc.org

indefinitely in the environment. Intentionally adding PFASs to food packaging leads to long-lasting environmental contamination and harmful exposures during manufacturing, use, and end-of-life.[3–5] PFASs can migrate from food packaging into food and be ingested.[4] After ingestion and excretion, PFASs end up in wastewater, where they often remain untreated, contaminating the environment and re-entering the food chain.[4] Composting plant fiber-based food packaging that contains PFASs can transfer PFASs from the packaging to the compost. For instance, Choi *et al.* (2019) demonstrated that composts made from municipal organic wastes containing food packaging had higher levels of PFASs than composts made from municipal organic wastes without food packaging.[6] Growing plants, especially food crops, in PFAS-containing compost may result in human exposure. Plant-fiber based food packaging that contains PFASs can also leach PFASs into the environment if landfilled, or can lead to accumulation of PFASs in the paper system if recycled.[5]

The use of PFASs in food packaging invariably leads to human exposures, with potential adverse health impacts (see Chapter 2 for a summary of known health effects). Additionally, PFAS food packaging additives contaminate the cycle of food production through multiple, interrelated pathways: from additives in direct contact with food, through indirect contamination of drinking water sources after food packaging disposal, and through contaminated soil, sediment, or compost used to grow new crops.[4]

PFASs in food packaging have faced increasingly stringent regulations in the United States (U.S.) over the last two decades, beginning with a 2002 EPA phase-out of some long-chain PFASs, which prompted food packaging producers to shift to shorter-chain PFASs, such as side-chain fluorinated polymers manufactured using 6:2 fluorotelomer alcohol (6:2 FTOH).[7,8] Reducing the chain length of the carbon-fluorine tails was assumed to confer the desired resistance to water and grease, while decreasing the hazards.[8] However, 6:2 FTOH degrades into highly persistent perfluorohexanoic acid (PFHxA) through toxic and biopersistent intermediate degradation products.[9,10]

In the U.S., several initiatives now limit the use of PFASs in food packaging. In 2020, the U.S. Food and Drug Administration (FDA) announced a voluntary phase-out of short-chain PFASs containing 6:2 FTOH.[11] As of 2020, the Biodegradable Products Institute, which certifies products to be biodegradable and thus suitable for composting facilities, no longer certifies any product that contains over 100 parts per million (ppm) of total fluorine.[12] As of December 2022, eleven states (California, Colorado, Connecticut, Hawaii, Maryland, Maine, Minnesota, New York, Rhode Island, Vermont, and Washington) have implemented bans on PFASs in food packaging.[13]

However, it is also important to acknowledge the potential trade-offs of such initiatives. While the original purpose of molded fiber food packaging products was to reduce plastic use, bans on plant fiber-based packaging products that use PFASs will most likely increase reliance on plastic food packaging once again. To prevent this return to widespread plastic usage, the food packaging industry needs a less hazardous alternative to PFASs with acceptable technical performance and good compatibility with current molded fiber manufacturing processes. Replacing PFASs with a benign and degradable alternative would greatly improve the product's end-of-life impacts, potentially closing the loop for molded fiber production (*e.g.*, if compost produced from molded fibers were used to grow trees for paper).

This project comprised a semester-long investigation of alternatives to PFAS-based product packaging as part of the UC Berkeley graduate course, Greener Solutions, from August to December 2020. Course instructors (Dr Megan Schwarzman, Mr Thomas McKeag, Dr William Hart-Cooper, and Ms Kimberly Hazard) guided the student team, comprising Minerva Teli (Environmental Engineering), Aaron Maruzzo (Environmental Health Sciences), Anna Kurianowicz (Biochemistry & Molecular Biology), and Kelly Chou (Chemistry). We describe the student team's process and findings below.

3.2 Approach

3.2.1 Project Scope

Fast food packaging products include a wide variety of materials and shapes made for foods with diverse barrier needs (Figure 3.1). This chapter focuses on plant fiber-based packaging materials, specifically molded fibers. Within the realm of paper packaging, existing reports on the transition away from PFASs tend to focus on 2D paper and paperboard packaging, recommending certain biopolymers such as polylactic acid (PLA) as replacements. However, many biopolymers have poor degradability in compost and reduce the quality of composting feedstock.[14] In contrast, 3D molded fiber materials are readily biodegradable and have good overall sustainability profiles. However, they have been consistently shown to contain high levels of PFASs,[15] which makes them an ideal focus for this chapter.

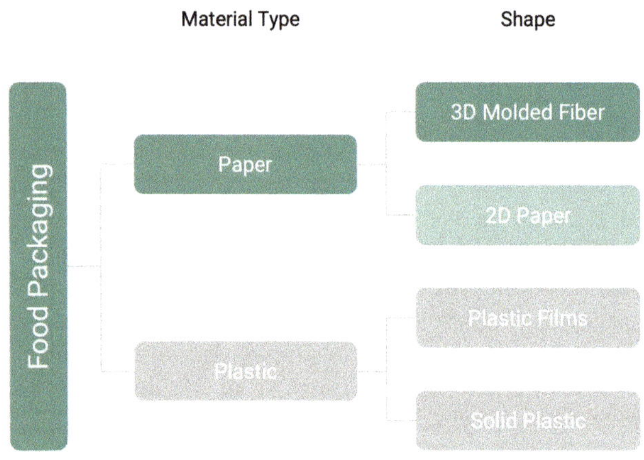

Figure 3.1 Fast food packaging can be divided into several broad categories. Our project focuses on 3D molded fiber packaging.

Molded fiber has many favorable qualities. It has a seamless shape that prevents leaks and can be customized based on the needs of the client. It can also be made from a recycled feedstock, similar to the transition from Step 9 to Step 5 in Figure 3.2, and the pulp fibers degrade readily in compost. However, the use of PFASs in molded fiber negates many of these positive attributes.

In a 2011 study conducted by toxic-free future, 100% of molded fiber products surveyed contained PFASs, and molded fiber products showed the highest levels of fluorine out of all items in the study.[15] Another study showed that pulp fiber materials were among the food contact materials

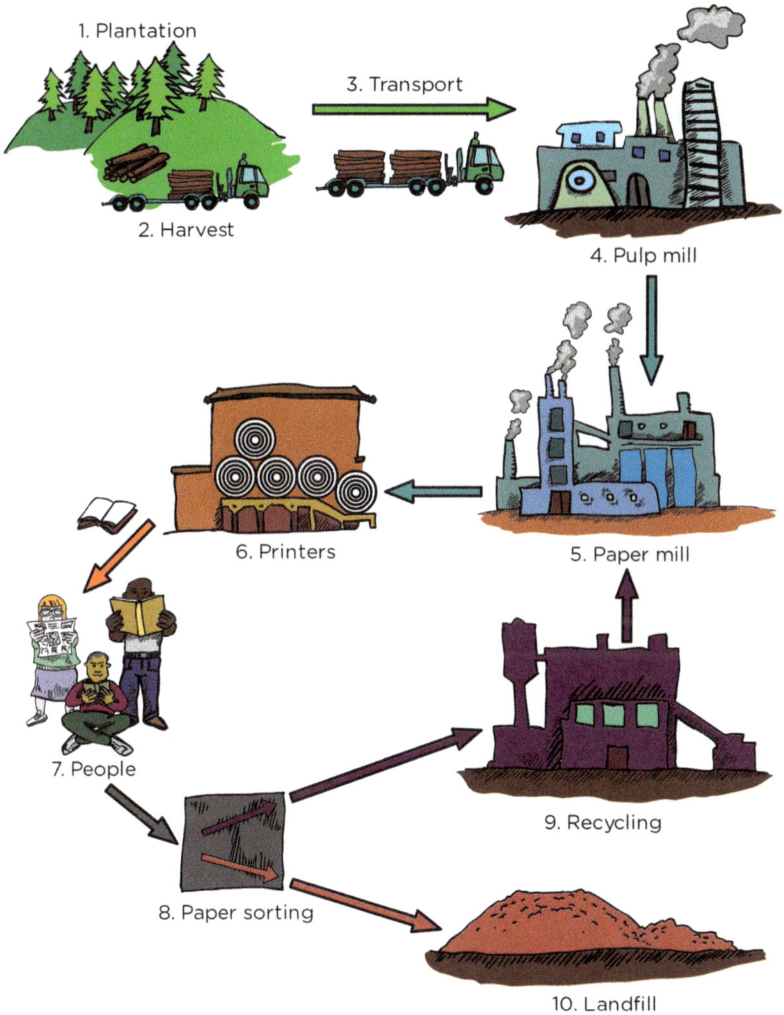

Figure 3.2 Schematic of paper production and life-cycle. Reproduced from ref. 76, 'Paper making process' by Siyavula Education, https://www.flickr.com/photos/121935927@N06/13579676055, under the terms of the CC BY 2.0 license, https://creativecommons.org/licenses/by/2.0/.

containing the highest levels of fluorotelomer alcohols (FTOHs).[16] The prevalence of PFASs in molded fibers is likely due to their specific manufacturing constraints. Paper-type products are generally produced as flat sheets, and then folded into a 3D shape, such as a cup, bag, or bowl. They can therefore be treated with a PFAS-free plastic lamination or spray coating while they are flat.[17] These factors make PFAS-free solutions more widely available for 2D paper products.

In contrast, molded fibers undergo the production process outlined in Figure 3.3, starting as bulk pulp fibers suspended in water in a blend tank, which are then molded and dried in an oven and packed for shipping. Molded fiber food packaging takes on a 3D, organic shape that is difficult to coat evenly once dried, rendering most dry-end processes prohibitively costly. Solutions applied in the form of films and sprays are difficult to adapt to a 3D molded fiber manufacturing process. Therefore, production of 3D molded fiber materials often relies exclusively on the addition of PFASs into the blend tank (known as "wet-end" treatment).[5] The major benefit of alternatives that directly replace PFASs as wet-end treatments in molded fiber manufacturing is that manufacturers do not have to execute major and costly changes to their machinery.

Another consideration is the sourcing of the alternative materials. New chemicals that need to be synthesized or extracted must be purchased from outside the molded fiber production system. The student team aimed to focus on alternative materials that are already part of the molded fiber production system, including materials that are considered waste or that are already being recycled and used in the system. This would enable the alternatives to be made in-house by molded fiber manufacturers, giving them even greater control over the physical properties of their product and reducing the environmental impact of transporting materials. This approach aims to overcome the challenge of chemical "lock-in", which forces manufacturers to depend on hazardous chemicals and materials because of the existing system of production.[18] Alternatives that directly replace PFASs and

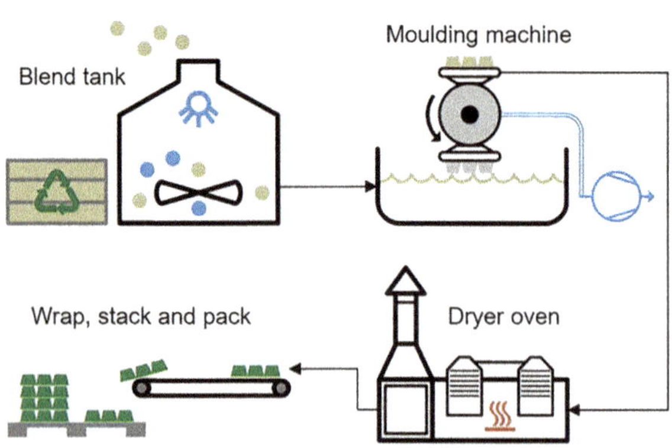

Figure 3.3 Schematic of the molded fiber production process. Reproduced from ref. 17 with permission from John Wiley & Sons, Ltd., Copyright 2017.

are more accessible to molded fiber manufacturers could help lower the barrier to switch to safer alternatives.

Thus, the student team aimed to identify accessible, safer, degradable chemicals that would be added to the wet-end of the molded fiber forming process to achieve water and oil repellency. The sections below describe the performance and hazard criteria that were considered and the materials that emerged as possible alternatives.

3.2.2 Technical Performance Criteria and Their Assessment

Table 3.1 presents the performance properties the student team used to evaluate the effectiveness of selected alternatives on molded fibers. In identifying these properties, the team considered the different needs of consumers, manufacturers, and composters (both facilities and at-home).

Consumers are primarily concerned with usability, which is strongly linked to issues of food leakage and preservation. This is best represented by barrier and surface properties. Three barrier and one surface properties are described below. The kit value measures the degree of repellency or anti-wicking of paper and boards.[19] This measurement was developed specifically

Table 3.1 Definitions of the properties evaluated in the performance assessment. Note that dispersibility and degradability are not industry standards, but rather, are based on the data available.

	Relevant property	Definition
Barrier properties	Grease resistance	Measured in kit value, a relative score of paper to absorb a series of oil-containing reagents
	Water contact angle	The angle between the surface of the liquid and the outline of the contact surface
	Water vapor permeability	The mass of water vapor at a certain pressure that permeates a material (with a certain thickness and surface area) over time
Surface properties	Porosity (oxygen permeability)	The volume of oxygen that permeates a material's surface area over time
Thermal properties	Thermal resistance	The temperature at which there is large drop in a material's weight relative to its original weight
	Temperature dependence	The temperature required for synthesis or formation of a material
Chemical properties	Dispersibility	Yes/No on whether the material can disperse in cellulose slurry
Mechanical properties	Tensile strength	The maximum load a material can support without fracture divided by its original cross-sectional area
End-of-life properties	Degradability	The amount of time for at least 70% of a material to break down in soil at ambient temperature containing no targeted microbes

for fluorine-containing papers and ranges from 1 to 12, with 12 indicating the highest level of repellency. The water contact angle indicates how well a droplet deposited on a solid substrate spreads out. If the droplet angle is above 90°, the material is considered hydrophobic; below 90°, the material is considered hydrophilic.[20] Water vapor permeability refers to water resistance in relation to humidity of the environment. A low value indicates better resistance and means the end product does not need an extremely dry environment for storage. The one surface property, porosity, indicates how close and strongly the fibers are bound together. Here it is measured in oxygen permeability, where a lower value indicates lower porosity and therefore a more densely packed material. A low oxygen permeability could also provide a slower rate of food decay when the product is in use.

In addition to barrier properties, manufacturers of fast food packaging products prioritize thermal, chemical, and mechanical properties. The thermal properties include thermal resistance and temperature dependence. Thermal resistance provides information on the final product's ability to withstand high temperatures. The data provided indicate the temperature at which the material completely breaks down, but it is important to note that the maximum temperature the material can withstand during food contact is lower. A higher temperature would partially compromise the product integrity, thus rendering the product less effective. The temperature dependence predicts whether the alternative can undergo any necessary reactions while withstanding temperatures up to 60 °C. This is because the alternative is mixed into the wet-end slurry at 60 °C.[17,21] The chemical property chosen by the student team is dispersibility, which is used to assess the potential for incorporation into the wet-end of processing. To be incorporated into molded fibers, the alternative must be able to evenly disperse in the cellulose slurry, with or without a dispersing agent. Mechanical properties are important to consider for transport, storage, and rigidity during use. A higher tensile strength allows the product to be compact during transport and carry a larger food load.

Finally, composters are primarily concerned with the end-of-life properties of the material. The student team chose degradability as the end-of-life property, indicative of compostability. These data were difficult to standardize, but the student team found that the included alternatives were able to degrade at least 70%.[14,22–24] While composting operations are often carried out at high temperatures or with targeted microorganisms, the student team considered degradability measurements at ambient soil temperatures with no targeted microbes. The purpose of this was not only to standardize the data, but also to help find alternatives that can be composted in more basic facilities or even in consumers' homes.

One novel strength of this assessment was the choice to use PLA as the performance baseline rather than molded fiber products containing PFASs. This selection was motivated in part by a recognition that PFASs overperform in imparting barrier properties. For example, PFOA on cellulose yields a water contact angle of 160°, whereas PLA only yields 75–85°.[25,26] Despite the difference in angle size, both consumers and manufacturers consider PLA as providing a sufficient barrier in typical storage scenarios. In other words, PLA is neither suboptimal nor optimal, but rather balances realistic performance expectations for molded fiber products with positive environmental attributes. In addition, the use of PLA as a baseline provides an opportunity to identify more sustainable alternatives. Although PLA is commonly regarded as compostable, it can only be composted by composting facilities that use high temperatures and it does not meet the home compostability standard.[14] Because the student team valued solutions that advance packaging sustainability, PLA as a technical baseline was not only realistic, but also advantageous.

Figure 3.4 lists all the performance data collected to compare the alternatives to PLA. Each row contains data on a single material from several references. It is important to note that each reference used the material under different conditions. Some were applied on a very thick layer of cellulose, while others were applied on cellulose with specific additives, and some were even used for an entirely different application. The data are representative of either the material itself, or of the material on cellulose (important for barrier and surface properties). The first column also indicates whether the sources applied the alternative as a film or integrated it within a fiber matrix. The performance of each alternative with regard to the different properties is indicated by the color of the boxes. Green indicates the best performance, yellow indicates moderate performance, red indicates worst performance, and gray signifies data gaps. These rankings are relative to one another, which means that a green box does not necessarily indicate the ideal option, but rather the best one among the selected alternatives.

3.2.3 Human and Environmental Health Criteria

The student team assessed fifteen hazard endpoints for each alternative: twelve human health endpoints and three environmental fate and toxicity endpoints (Figures 3.5, 3.8, 3.11, and 3.13).

For each candidate alternative, hazard information was collected from authoritative listings from Pharos, data from the European Chemicals Agency, primary literature, or predictive models from U.S. EPA CompTox or ToxTree (see Figure 3.11). Endpoint data not provided by these sources were collected from Material Safety Data Sheets or agency administrative reports, or were inferred by the student team based on 12 physicochemical characteristics collected from PubChem. Each hazard score (from very high (vH) to low (L) or very low (vL)) was assigned a confidence level based on the reliability, consistency, and completeness of information sources. A high confidence level indicates that the data come from authoritative lists or assays compliant with the principles of Good Laboratory Practice, and a low confidence level indicates that the measured data were conflicting or incomplete, originating solely from predictive models, or were inferred by the student team.

	Application of Data Source	Barrier Properties			Surface Properties	Thermal Properties		Chemical Properties	Mechanical Properties	End-of-Life Properties
Relevant Property	film or within molded fiber matrix	grease resistance	water contact angle	water vapor permeability (WVP)	porosity (oxygen permeability)	thermal resistance (temp at major % weight loss)	temperature dependence (manufacture temp 60°C)	dispersibility	tensile strength	degradability (at least 70% in soil)
Units	--	kit value	degrees	g.mm/m2.kPa.day	cm3/m2/24hr	°C	°C	--	MPa	days
Polylactic Acid	Films	12+[27]	75 - 85[26]	0.161-0.237[27]	38-42[28]	335°C[29]	52.5-150°C[26]	DG	27-35[30]	> 150[14]
Cellulose Nanocrystals	Films	1131	6531	932	DG	250°C[33]	30-60°C[33]	Yes[33]	15.633	224
Lignin	Fiber Matrix and Films	1234	8935	5.3 ± 0.1[34]	26.6 ± 0.5[34]	260-290°C[35]	85°C[35]	Yes, with agent[35]	20.336	2323

Key				
	High Efficiency	Moderate Efficiency	Low Efficiency	Data Gap (DG)

Figure 3.4 Technical assessment of PLA[14,26–30] and two selected alternatives (cellulose nanocrystals[24,31–33] and lignin[23,34–36]).

	Chemical Name	C	M	R	D	E	AT	ST*	N	SnS*	SnR*	IrS	IrE	AA/ CA	P	B
Baseline Chemicals	6:2 fluorotelomer methacrylate	*M*[37]	*L*[38]	**M**[38]	**M**[38]	*M*[a]	*H*	**M**[38]	DG	*L*[38]	*L*[a]	**H**[38]	**H**[38]	*H*[a,38]	*vH*[a,38]	*L*[38]
	6:2 fluorotelomer alcohol	*M*[39,43]	*L*[40]	**L**[41]	*M*[41,43]	*M*[42]	**H**[41]	**vH**[41]	DG	**L**[40,43]	*L*[43]	**L**[40,43]	**L**[40,43]	*H*[41]	**vH**[41]	*L*[41]
	2-(N,N-Diethylamino)ethyl methacrylate	*L*[37]	*L*[37]	*H*[44]	*M*[45]	DG	**H**[45]	DG	DG	**H**[45]	DG	**vH**[45]	*vH*[45]	*H*[44]	**vL**[46]	**vL**[46]

Key					
	Hazard	High (H) / Very High (VH)	Moderate (M)	Low (L)	Data Gap (DG)
	Confidence	*Low*	**High**		

Figure 3.5 Hazard table for three baseline bad actor chemicals: 6:2 fluorotelomer methacrylate,[37,38] 6:2 fluorotelomer alcohol,[39–43] and 2-(*N*,*N*-diethylamino)ethyl methacrylate.[37,44–46]

Figure 3.5 lists the hazards associated with the baseline bad actor chemicals present in molded fiber manufacturing. The student team selected two PFASs and one non-fluorinated methyacrylate to represent the mixtures of PFASs present in molded fiber manufacturing: 6:2 FTOH (CAS no. 678-39-8), tridecafluorohexylethyl methacrylate (6:2 FTMAC, CAS no. 2144-53-8), and 2-*N*,*N*-diethylamino)ethyl methacrylate (CAS no. 105-16-8). At the time of the report, more than half of 17 distinct formulations approved for use in food packaging in the U.S. were polymer mixtures that contained a 6:2 FTMAC copolymer, and seven formulations contained an acrylate copolymer.[47,48] 6:2 FTOH is the parent compound of 6:2 FTMAC and its intermediate degradation product commonly detected as an impurity in food packaging surveys. Although 2-(*N*,*N*-diethylamino)ethyl methacrylate is not a PFAS, it was included in the baseline group because food packaging surface treatments are not composed exclusively of PFASs, but instead are often mixtures of PFASs and a methacrylate.[47] Its inclusion aimed to better represent the hazards that originate from a mixture that may have been used in molded fiber packaging. The minimum criteria to fulfil the definition of "safer" in this assessment were as follows: (1) reduced group I or group II/II* endpoints[†] compared to the baseline substances, and (2) minimal environmental toxicity and low levels of persistence and bioaccumulation.

3.3 Alternative Solutions

3.3.1 Cellulose Nanocrystals (CNCs)

3.3.1.1 Inspiration for CNCs

Cellulose, in its bulk form, is the most abundant natural polymer on Earth and the primary material used to make paper products. In its natural form, cellulose consists of both amorphous and crystalline regions.[49] When the crystalline regions of natural cellulose are isolated and recrystallized on the nano-scale (Figure 3.6), cellulose acquires hydrophobic properties without loss of degradability.[50] Cellulose nanocrystals (CNCs) are widely studied as stabilizers for improving structural stability of bio-composite films.[50–52] They can be made from a wide variety of plant sources and have many desirable properties when used as films.

3.3.1.2 Method of Incorporation

Bulk cellulose fiber or paper pulp is first digested in acid to hydrolyze bonds in natural cellulose, deconstruct the amorphous regions of the polymer, and

[†]Group I endpoints include carcinogenicity (C), mutagenicity (M), reproductive toxicity (R), development toxicity (D), and endocrine toxicity (E), and group II/II* endpoints include acute mammalian toxicity (AT), systemic toxicity and organ effects (ST*), neurotoxicity (N), skin sensitization (SnS*), respiratory sensitization (SnR*), skin irritation (IrS), and eye irritation (IrE). *Repeated exposure.

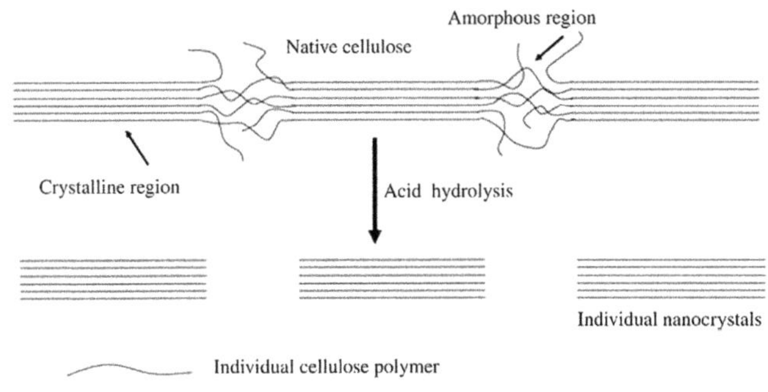

Figure 3.6 Process of transforming natural cellulose into cellulose nanocrystals. Reproduced from ref. 77 with permission from Elsevier, Copyright 2017.

form CNCs.[33] These CNCs have the same chemical composition as natural cellulose, but exhibit a more rigid and hydrophobic crystal structure.[32,49] Following production, CNCs can be directly added to the pulp fiber blend tank. While the dispersibility of CNCs depends on the polarity and purity of the mixture, which vary among manufacturers, some reports have shown successful incorporation into films without any added dispersing agent.[33] The CNCs could be produced in-house by molded fiber manufacturers from the same bulk fibers used in molded fibers. This method has great potential, especially for manufacturers that use recycled paper for their pulp feedstock. Not only can CNCs be prepared using recycled paper, they also add stability to shorter fibers, the presence of which is often a disadvantage of using recycled paper materials.[52] In-house production of CNCs by manufacturers could cut costs of purchasing external additives, improve sustainability from reduced transport of materials, and provide control over the specific properties of the CNCs. However, this would require investment into the infrastructure to support a new acid-digestion processing step. In the interim, CNCs could still be purchased from an external supplier.

3.3.1.2.1 Technical Performance. Technical data available for CNCs are largely for film systems (Figure 3.7), so the values may differ when incorporated into a fiber matrix. Nonetheless, relative to PLA, CNC films perform well in the grease resistance category, having a kit value comparable to that of PLA and a moderate water contact angle.[53,54] However, in terms of water vapor permeability, CNCs perform much worse than PLA. The value of 9 g mm per m^2 per kPa per day indicates that a significant amount of water vapor can escape from the film,[32] although this value is also strongly influenced by the type of film used in the study. Incorporation of CNCs into the molded fiber matrix shows promise based on its dispersibility but may be of concern during wet-end processing due its lower temperature dependence. If CNCs degrade slightly at the high temperature of the cellulose slurry, this may affect the mechanical properties of the final molded fiber product.

	Application of Data Source	Barrier Properties			Surface Properties	Thermal Properties		Chemical Properties	Mechanical Properties	End-of-Life Properties
Relevant Property	film or within molded fiber matrix	grease resistance	water contact angle	water vapor permeability (WVP)	porosity (oxygen permeability)	thermal resistance (temp at major % weight loss)	temperature dependence (manufacture temp 60˚C)	dispersibility	tensile strength	degradability (at least 70% in soil)
Units	--	kit value	degrees	g.mm/m2.k Pa.day	cm3/m2/24hr	˚C	˚C	--	MPa	days
Polylactic Acid	Films	12+	75 - 85	0.161-0.237	38-42	335˚C	52.5-150˚C	DG	27-35	> 150
Cellulose Nanocrystals	Films	11	65	9	DG	250˚C	30-60˚C	Yes	15.6	2

Key	Relative Rankings	Best	Moderate	Worst	Data Gap (DG)

Figure 3.7 Technical performance assessment of CNCs.

3.3.1.2.2 Health and Environmental Criteria. CNCs lacked hazard information from authoritative lists at the time of this project. Despite this, a preliminary hazard assessment for wood-derived CNCs was performed based on expert judgment and peer-reviewed toxicity studies from Shatkin *et al.* (2016),[55] Roman (2015),[61] and Shatkin and Kim (2015)[62] (Figure 3.8).

The parent compound of CNCs is cellulose, which is benign overall; however, the acid treatment and resulting production of nanoscale particles slightly shift the hazard profiles of CNCs. The particle dimensions of CNCs are 130 ± 67 nm in length and 5.9 ± 1.8 nm in width, which results in a high surface area to volume ratio.[62] A hazard comparison between CNCs and baseline chemicals indicates that CNCs would improve the relative safety of most known group I and II/II* endpoints by an average of one order of magnitude, except for respiratory sensitization. One occupational epidemiology study identified an association with reduced lung function following repeated exposures.[62]

According to Shatkin *et al.* (2015), there is sufficient information about the environmental attributes of CNCs, and aquatic toxicity assays in zebrafish indicate low toxicity.[62] CNCs are not known or expected to persist in the environment. However, their bioaccumulation potential is unknown, and existing predictive models cannot be used to estimate this endpoint. A better approximation of CNC bioaccumulation would entail the adaptation of traditional predictive methods and physicochemical characterization, which were beyond the scope of the hazard assessment.[63]

Finally, CNC manufacture and incorporation into food packaging materials relies on dissolving cellulose fibers in a strong acid, such as sulfuric acid. A preliminary hazard assessment of sulfuric acid suggests a potentially concerning hazard profile compared to the baseline. While its primary human health hazards, including carcinogenicity and mutagenicity, are lower by one order of magnitude, sulfuric acid's secondary human health endpoints such as mammalian toxicity and irritation are equivalent to or worse than those of the baseline chemicals.[58,59] However, caution in interpretation is recommended here. These chemicals are representative of the manufacturing process required for this strategy but are not expected to be present in any meaningful amount in the final molded fiber product. The resulting partial esterification of the hydroxyl groups on the CNCs results in a slight acidity, although this acidity may be considered minimal against the buffering capacities of mammalian organisms.[61] The process of cellulose acidification already exists as a hazard in the papermaking process, including molded fiber manufacturing, that presumably can be minimized using effective management controls.

3.3.2 Lignin

3.3.2.1 Inspiration for Lignin

Lignin is a biopolymer that makes up 15–40% of plant biomass, depending on the species of plants. It is most abundant in the cell walls and is the primary source of cell wall rigidity.[64] Lignin is removed as a waste product when

	Chemical Name	C	M	R	D	E	AT	ST*	N	SnS*	SnR*	IrS	IrE	AA/CA	P	B
Baseline Chemicals	6:2 fluorotelomer methacrylate	*M*	*L*	**M**	**M**	*M*[a]	*H*	**M**	DG	*L*	*L*[a]	**H**	**H**	*H*[a]	*vH*[a]	*L*
	6:2 fluorotelomer alcohol	*M*	*L*	**L**	*M*	*M*	**H**	**vH**	DG	**L**	*L*	**L**	**L**	*H*	**vH**	*L*
	2-(N,N-Diethylamino)ethyl methacrylate	*L*	*L*	*H*	*M*	DG	**H**	DG	DG	**H**	DG	**vH**	*vH*	*H*	**vL**	**vL**
CNC Alternative	Nanocellulose crystals (CNC)	*L*[55]	**L**[55]	DG	DG	DG	**L**[55]	*M*[55]	DG	**L**[56]	*M*[55,61]	**L**[56]	*M*[55]	*L*[55]	*vL*[57]	*vL*[55,57]
	Sulfuric acid	**M**[58]	*L*[59]	*L*[59]	*M*[59,60]	DG	*vH*[59]	*M*[59]	DG	*L*[59]	**H**[60]	**vH**[58]	**vH**[58]	*M*[60]	*vH*[60]	**vL**[59]

Key

Hazard	High (H)	Moderate (M)	Low (L)	Data Gap (DG)

Figure 3.8 Summary hazard table for CNCs[55–57] and sulfuric acid.[58–60] [a]Read-across from 6:2 FTOH (degradation product of 6:2 FTMAC).

cellulose is isolated from raw wood chips during the papermaking process, due to its insolubility in the cellulose slurry and its ability to clump cellulose fibers together through cross-linking. The paper industry produces 70 million tons of lignin waste per year.[64] Lignin is also a common waste product in other industries, such as ethanol production. Once isolated, lignin forms a black liquor, which is often used as an internal energy input for the pulping process.[64] Alternatively, this biopolymer could provide barrier and structural integrity to the molded fiber matrix. Since lignin is known to bind well to cellulose and provide structure to plants, it makes sense that this waste product could both be well incorporated into and provide rigidity to molded fibers.

3.3.2.2 *Method of Incorporation*

To be incorporated into the molded fiber, lignin must be extracted from the black liquor using carbon dioxide and acid treatment (Figure 3.9). Traditionally, this process uses a strong acid like sulfuric acid, which is also utilized during the stripping process in papermaking. However, this aggressive chemical could be replaced with non-toxic chemicals such as potassium aluminum sulfate.[65] Lewis *et al.* (2016)[35] reported that when adding lignin to the molded fiber wet-end slurry, the addition of an ionic liquid helps suspend the lignin, dispersing it more evenly and preventing fiber clumping. Incorporation of other plant materials, such as xylan, can improve molded fiber porosity.[35]

Another method of incorporating lignin into molded fibers is through light delignification. As mentioned earlier, lignin is extracted from raw materials to form a cellulose slurry during papermaking. Instead of fully removing the lignin content from the raw material, some of the lignin can be

Figure 3.9 Lignin extraction process from black liquor. Reproduced from ref. 65 with permission from Springer Nature, Copyright 2012.

retained, so the bonds between the original lignin and cellulose are preserved. It is important to note that this process is only possible if the molded fiber is manufactured from virgin material as opposed to recycled paper. Molded fiber production from light-delignified cellulose has been successfully achieved by Wang *et al.*,[66] although not in the context of fast food packaging.

3.3.2.2.1 Technical Performance. The data presented in Figure 3.10 come primarily from sources that apply lignin within the molded fiber matrix. Compared to CNCs, lignin has better technical performance. Under barrier properties, the kit value and water contact angle are comparable to those of PLA.[35,67] In fact, the water contact angle is slightly higher, meaning lignin is slightly more hydrophobic. The water vapor permeability is significantly higher than that for PLA, which means that the final product may require less humid storage conditions.[67] Lignin has a slightly lower porosity than PLA,[67] but combining it with cellulose will most likely result in a higher porosity than PLA.

Its thermal resistance value indicates that lignin is suitable for hot food but breaks down faster than PLA when comes in contact with heat. Its temperature during manufacture appears to be compatible with the molded fiber pulp slurry, so there is little concern that the lignin will break down at the wet-end, and additional heat will not be necessary. As mentioned earlier, lignin normally causes clumping when added to the cellulose slurry; however if it is added back with a dispersing agent, such as an ionic liquid, it can be evenly distributed. Finally, the tensile strength of lignin is very close to that of PLA and its degradability is significantly improved. Although lignin takes the longest time among our alternatives to degrade, this time shortens when it is exposed to fungi and bacteria commonly found in temperate soils.[68]

3.3.2.2.2 Health and Environmental Criteria. The U.S. EPA evaluated lignin under the Safer Choice program and listed it on the Safer Chemical Ingredient List after the verification of low concern through experimental and modeled data.[71] Additional studies identified lignin as an anti-oxidant, exhibiting medicinal roles against non-communicable diseases such as cancer and obesity.[72] Accordingly, the hazard assessment shown in Figure 3.11 categorizes lignin as benign, with all hazards in each endpoint minimized compared to the baseline.

The hazards of three additional chemicals related to the process of using lignin were also evaluated. Xylan is a hemicellulose conferring tensile strength, sulfuric acid is an existing chemical used in the papermaking process and to recover lignin waste, and the ionic liquid 1-allyl-3-methylimidazolium is a potential dispersing agent used to prepare and re-incorporate lignin. Both sulfuric acid and the ionic liquid have potentially concerning hazard profiles. However, as mentioned for the CNC alternative, sulfuric acid is an existing chemical within the papermaking industry with relatively fewer occupational risks compared to other industry sectors due to minimal workplace exposure and appropriate workplace controls. The hazards associated with 1-allyl-3-methylimidazolium are lower than the PFAS baseline for carcinogenicity,

	Application of Data Source	Barrier Properties			Surface Properties	Thermal Properties		Chemical Properties	Mechanical Properties	End-of-Life Properties
Relevant Property	film or within molded fiber matrix	grease resistance	water contact angle	water vapor permeability (WVP)	porosity (oxygen permeability)	thermal resistance (temp at major % weight loss)	temperature dependence (manufacture temp 60°C)	dispersibility	tensile strength	degradability (at least 70% in soil)
Units	--	kit value	degrees	g.mm/m2.kPa.day	cm3/m2/24hr	°C	°C	--	MPa	days
Polylactic Acid	Films	12+	75 - 85	0.161-0.237	38-42	335°C	52.5-150°C	DG	27-35	> 150
Lignin	Fiber Matrix and Films	12	89	5.3 ± 0.1	26.6 ± 0.5	260-290°C	85°C	Yes, with agent	20.3	23

Key	Relative Rankings	Best	Moderate	Worst	Data Gap (DG)

Figure 3.10 Technical performance assessment of lignin.

	Chemical Name	C	M	R	D	E	AT	ST*	N	SnS*	SnR*	IrS	IrE	AA/CA	P	B
Baseline Chemicals	6:2 fluorotelomer methacrylate	*M*	*L*	**M**	**M**	*M*[a]	*H*	**M**	DG	*L*	*L*[a]	**H**	**H**	*H*[a]	*vH*[a]	*L*
	6:2 fluorotelomer alcohol	*M*	*L*	**L**	*M*	*M*	**H**	**vH**	DG	**L**	*L*	**L**	**L**	*H*	**vH**	*L*
	2-(N,N-Diethylamino)ethyl methacrylate	*L*	*L*	*H*	*M*	DG	**H**	DG	DG	**H**	DG	**vH**	*vH*	*H*	**vL**	**vL**
Lignin Alternative	Lignin															
	Xylan															
	Sulfuric acid	**M**[58]	*L*[59]	*L*[59]	*M*[59,60]	DG	*vH*[59]	*M*[59]	DG	*L*[59]	**H**[60]	**vH**[58]	**vH**[58]	*M*[60]	*vH*[60]	**vL**[59]
	1-allyl-3-methylimidazolium chloride	*L*[37]	*L*[b,69]	DG	DG	*L*[46]	*M*[b,71]	**M**[70]	DG	*L*[37]	DG	**H**[70]	**H**[70]	DG	*L*[46]	*vL*[46]

Key

Hazard	High (H)	Moderate (M)	Low (L)	Data Gap (DG)

Figure 3.11 Summary hazard table for lignin and associated process chemicals (xylan, sulfuric acid, and 1-allyl-3-methylimidazolium chloride[37,46,69,70]).

endocrine toxicity, acute mammalian toxicity, and persistence.[42,46,71] However, 1-allyl-3-methylimidazolium scored similar to PFASs for irritation endpoints, which may be of concern in occupational settings but could be mitigated with appropriate controls.[70]

Lignin is a naturally-occurring component of soil organic matter and outperforms PFASs in terms of environmental hazards.[68,73] Thus, the addition of lignin to molded fibers may be one of the safest and practical alternative strategies available.

3.4 Conclusion

3.4.1 Technical Comparison

3.4.1.1 Molded Fiber Incorporation

Lignin had technical data available regarding its incorporation into the fiber matrix, whereas CNCs had data primarily focused on their application as a film on the surface. Including dispersibility and temperature dependence into the technical assessment attempts to address whether the CNC film could be applied within a cellulose fiber matrix. In terms of dispersibility, both alternatives show promise of spreading evenly in the wet-end slurry, although lignin requires an additive to achieve this. The temperature of the slurry may be of concern for CNCs, since they are synthesized at a lower or equivalent temperature than that of the slurry and may break down before reaching the dry-end phase. Further research into applying CNCs in a cellulose fiber matrix is required.

3.4.1.2 Food Type

Figure 3.12 displays three types of foods that are commonly stored in molded fibers: room temperature foods with low barrier needs (*e.g.*, salad), hot and oil-heavy foods (*e.g.*, fries and burgers), and hot foods with moderate oil and water content (*e.g.*, pasta). Based on the technical assessments, the alternatives are categorized based on the food types they are most suitable for serving.

Both alternatives could possibly be used in molded fiber salad bowls, because of the low temperature and limited barriers requirements. Both are also suitable for use in packing greasier foods, since their kit values are very close to that of PLA and they display medium levels of thermal resistance. The high thermal resistance of PLA might not be necessary at the temperature of hot foods. This high resistance most likely overperforms for its purpose and may contribute to PLA's long degradation time.

The high temperature and medium barrier requirements for foods like pasta match well with the technical capabilities of lignin, but not CNCs. Lignin has a high water contact angle and breaks down at relatively high temperatures. These foods also tend to be heavier, which is suitable for lignin since its tensile strength is very close to that of PLA.

Possible Alternatives	Nanocellulose	Lignin	Nanocellulose	Lignin	Lignin
Food Type					
Temperature	Room Temperature		Hot		Hot
Water Content	Medium		Low		Medium
Oil Content	Low		High		Medium

Figure 3.12 Assignment of the food type for each alternative based on the technical performance assessment. Reproduced from ref. 78, 'Curry Chickpea Bowl @sweetgreen' by Shashi Bellamkonda, https://www.flickr.com/photos/drbeachvacation/32574005580/, ref. 79, 'Burger and fries' by Alice Salles, https://www.flickr.com/photos/132313505@N07/20554610314/, and ref. 80 'Cherry tomatoes and pasta' by jeffreyw, https://www.flickr.com/photos/jeffreyww/14530170849/, under the terms of the CC BY 2.0 license, https://creativecommons.org/licenses/by/2.0/.

3.4.2 Sustainability

A large emphasis in this project was placed on finding alternatives that could be sustainably sourced and disposed of. This led to emphasizing sourcing within the molded fiber manufacturing industry and including the degradation criteria, respectively. As demonstrated in the degradability section of the technical assessment, both alternatives are capable of decomposing within a month, without high temperatures or added microbes. This shows great promise that CNCs and lignin could be composted at virtually any composting facility and likely also in a home setting.

A major benefit of lignin and CNCs is that they can be sourced from within the paper industry, meaning less processing energy and materials. Lignin does require processing from its waste form, but by utilizing a waste product, little new material is needed. CNCs introduce no new materials but do involve extra processing that creates more acidic waste. Note that the acid required for both these processes is already used in the papermaking process, so it is likely accessible to manufacturers.

3.4.3 Hazard Comparison

The candidate alternative chemicals exhibited significant improvements in safety across known group I endpoints with orders of magnitude differences (Figure 3.13). Most group II/II* endpoints, such as acute and systemic toxicity, skin sensitization, respiratory sensitization, and some irritation

	Chemical Name	C	M	R	D	E	AT	ST*	N	SnS*	SnR*	IrS	IrE	AA/CA	P	B
Baseline Chemicals	6:2 fluorotelomer methacrylate	*M*	*L*	**M**	**M**	*M*[a]	*H*	**M**	DG	*L*	*L*[a]	**H**	**H**	*H*[a]	*vH*[a]	*L*
	6:2 fluorotelomer alcohol	*M*	*L*	**L**	*M*	*M*	**H**	**vH**	DG	**L**	*L*	**L**	**L**	*H*	**vH**	*L*
	2-(N,N-Diethylamino)ethyl methacrylate	*L*	*L*	*H*	*M*	DG	**H**	DG	DG	**H**	DG	**vH**	*vH*	*H*	**vL**	**vL**
CNC Alternative	Nanocellulose crystals (CNC)	*L*	**L**	DG	DG	DG	**L**	*M*	DG	**L**	*M*	**L**	*M*	*L*	*vL*	*vL*
	Sulfuric acid	**M**	*L*	*L*	*M*	DG	*vH*	*M*	DG	*L*	**H**	**vH**	**vH**	*M*	*vH*	**vL**
Lignin Alternative	Lignin															
	Xylan															
	Sulfuric acid	**M**	*L*	*L*	*M*	DG	*vH*	*M*	DG	*L*	**H**	**vH**	**vH**	*M*	*vH*	**vL**
	1-allyl-3-methylimidazolium chloride	**L**	*L*[b]	*DG*	*DG*	L	*M*[b]	*M*	DG	*L*	**DG**	**H**	**H**	*DG*	*L*	**vL**

Key					
	Hazard	High (H) / Very High (VH)	Moderate (M)	Low (L)	Data Gap (DG)
	Confidence	*Low*	**High**		

Figure 3.13 Summary hazard table of alternative strategies. [a] Read-across from 6:2 FTOH (degradation product of 6:2 FTMAC). [b] Read-across from 1-ethyl-3-methylimidazolium chloride (89% structurally similar to 1-allyl-3-methylimidazolium chloride according to EPA Comp Tox).

endpoints, were also reduced compared to the baseline. However, the hazards posed by some process chemicals were slightly higher than baseline. For example, scores for the eye or skin irritation endpoints were equivalent or worse compared to the existing chemicals.

Neurotoxicity and endocrine toxicity were notable data gaps, with limited to no toxicity data available for evaluation in either baseline chemicals or alternatives. Data gaps and uncertainties, primarily for CNCs, point to opportunities for further research. The student team used *in silico* methods to detect structural similarities with known carcinogens, mutagens, sensitizers, persistent chemicals, and bioaccumulative chemicals, but more work is needed to corroborate results and interpret them.

3.4.4 Next Steps

To continue the search for safer alternatives to PFASs in molded fiber products, we identified three broad next steps for researchers and molded fiber manufacturers:

3.4.4.1 Continue Research and Testing of Alternatives in a Molded Fiber Context

Since both alternatives are novel, they have not been tested in molded fibers, but are worth investigating. The strategies listed here (CNCs and lignin) may also be combined synergistically, for example, development of lignin-containing nanocellulose[74] and incorporation of lignin nanoparticles and CNCs into polyvinyl alcohol films.[75] Also, toxicity studies linked to the oral ingestion route are necessary if CNCs are adopted as an alternative and come in contact with food.

3.4.4.2 Create Common Language and Better Transparency in the Food Packaging Industry

A need for this became apparent given the challenges that surfaced during the technical assessments—available data were extremely limited and varied wildly in terms of the parameters tested, testing methods, and base materials used, causing discrepancies and data gaps. Of the available data, some technical performance parameters were standardized (*e.g.*, water contact angle), while other metrics for performance, such as oil resistance and porosity, were measured using a variety of tests that could not be compared.

3.4.4.3 Explore Opportunities for Further Green Innovation

Green innovation does not end with replacing PFASs. The paper and molded fiber production process involves a host of other materials and treatments.

Traditional manufacturing chemicals like strong acids could be substituted through a similar iterative search for chemical alternatives *via* performance and hazard assessments.

Safer alternatives to PFASs in molded fiber products can ensure that fiber-based food packaging products remain viable plastic replacements. The availability of chemically safe and high-performing molded fiber packaging may help decrease the use of plastic, with long-term benefits to human health and the environment.

Abbreviations

AA/CA	Acute and chronic aquatic toxicity
AT	Acute mammalian toxicity
B	Bioaccumulation
C	Carcinogenicity
CNC	Cellulose nanocrystals
D	Developmental toxicity
E	Endocrine toxicity
ECHA	European Chemical Agency
EPA	Environmental Protection Agency
FDA	Food and Drug Administration
FTMAC	Fluorotelomer methacrylate
FTOH	Fluorotelomer alcohol
IrE	Eye irritation
IrS	Skin irritation
M	Mutagenicity
N	Neurotoxicity
P	Persistence
PFASs	Perfluoroalkyl and polyfluoroalkyl substances
PLA	Polylactic acid
R	Reproductive toxicity
SnR*	Respiratory sensitization
SnS*	Skin sensitization
ST*	Systemic toxicity and organ effects (incl. immunotoxicity)

Acknowledgements

This chapter is adapted from a student report originally written by Minerva Teli, Aaron Maruzzo, Kelly Chou, and Anna Kurianowicz. The authors express their gratitude to the Greener Solutions, Fall 2020 teaching team at the University of California, Berkeley: Meg Schwarzman, Tom McKeag, Billy Hart-Cooper, and Kim Hazard for their insightful feedback and mentorship in their search for greener solutions. The authors are thankful to the industry mentor, Kaj Johnson, for his guidance and support throughout this process. The authors would also like to extend their thanks to the food packaging experts Greg Glenn, Lauren Olsen, and Marty Mulvihill, who helped in defining the scope and gave important insights into the industry needs. Finally, the authors would like to thank Amanda Guan, Tom Bruton, and Kim Hazard, who were vital in transforming the original student report into the chapter it has become.

References

1. The False Promise Of Bioplastics and Compostable Plastics, https://www.beyondplastics.org/fact-sheets/bad-news-about-bioplastics (accessed 26 January 2023).
2. Z. Wang, J. C. Dewitt, C. P. Higgins and I. T. Cousins, *Environ. Sci. Technol.*, 2017, **51**, 2508–2518.
3. L. A. Schaider, S. A. Balan, A. Blum, D. Q. Andrews, M. J. Strynar, M. E. Dickinson, D. M. Lunderberg, J. R. Lang and G. F. Peaslee, *Environ. Sci. Technol. Lett.*, 2017, **4**, 105–111.
4. A. Ramírez Carnero, A. Lestido-Cardama, P. Vazquez Loureiro, L. Barbosa-Pereira, A. Rodríguez Bernaldo de Quirós and R. Sendón, *Foods*, 2021, **10**, 1443.
5. G. Glenn, R. Shogren, X. Jin, W. Orts, W. Hart-Cooper and L. Olson, *Compr. Rev. Food Sci. Food Saf.*, 2021, **20**, 2596–2625.
6. Y. J. Choi, R. Kim Lazcano, P. Yousefi, H. Trim and L. S. Lee, *Environ. Sci. Technol. Lett.*, 2019, **6**, 372–377.
7. U.S. Environmental Protection Agency, *Long-Chain Perfluorinated Chemicals (PFCs) Action Plan*, 2009.
8. C. F. Kwiatkowski, D. Q. Andrews, L. S. Birnbaum, T. A. Bruton, J. C. Dewitt, D. R. U. Knappe, M. V. Maffini, M. F. Miller, K. E. Pelch, A. Reade, A. Soehl, X. Trier, M. Venier, C. C. Wagner, Z. Wang and A. Blum, *Environ. Sci. Technol. Lett.*, 2020, **7**, 532–543.
9. S. V. Kabadi, J. W. Fisher, D. R. Doerge, D. Mehta, J. Aungst and P. Rice, *Toxicol. Appl. Pharmacol.*, 2020, **388**, 114878.
10. P. A. Rice, J. Aungst, J. Cooper, O. Bandele and S. V. Kabadi, *Food Chem. Toxicol.*, 2020, **138**, 111210.
11. Center for Food Safety and Applied Nutrition, FDA Announces the Voluntary Phase-Out by Industry of Certain PFAS Used in Food Packaging, https://www.fda.gov/food/cfsan-constituent-updates/fda-announces-voluntary-phase-out-industry-certain-pfas-used-food-packaging (accessed 26 January 2023).
12. Biodegradable Products Institute – Fluorinated Chemicals, https://bpiworld.org/Fluorinated-Chemicals (accessed 22 September 2022).
13. J. S. Baughan, J. Dages, J. Cooper and C. Rolph, State of the States, https://www.steptoe.com/en/news-publications/state-of-the-states-pfas-food-packaging-bans-and-restrictions.html (accessed 19 February 2023).
14. C. Vasile, D. Pamfil, M. Râpă, R. N. Darie-Niţă, A. C. Mitelut, E. E. Popa, P. A. Popescu, M. C. Draghici and M. E. Popa, *Composites, Part B*, 2018, **142**, 251–262.
15. Packaged in Pollution, https://toxicfreefuture.org/research/packaged-in-pollution/ (accessed 6 November 2022).

16. G. Yuan, H. Peng, C. Huang and J. Hu, *Environ. Sci. Technol.*, 2016, **50**, 942–950.
17. M. Didone, P. Saxena, E. Brilhuis-Meijer, G. Tosello, G. Bissacco, T. C. Mcaloone, D. C. A. Pigosso and T. J. Howard, *Packag. Technol. Sci.*, 2017, **30**, 231–249.
18. J. Blumenthal, M. L. Diamond, M. Hoffmann and Z. Wang, *Environ. Sci. Technol.*, 2022, **56**(7), 3863–3870.
19. Grease Resistance|Paper Testing Physical Properties, https://www.smithers.com/industries/packaging/manufacturers-and-users/packaging-materials-testing/paper-testing-surface-properties/grease-resistance (accessed 22 September 2022).
20. Contact angle, https://www.kruss-scientific.com/en/know-how/glossary/contact-angle (accessed 22 September 2022).
21. J. Hyde, Fybrel® for Molded Fiber Applications Producing High Strength Molded Fiber Products, https://www.minifibers.com/documents/Fybrel-for-Molded-Fiber-Applications.pdf (accessed 29 January 2023).
22. P. Singh and V. P. Sharma, *Procedia Environ. Sci.*, 2016, **35**, 692–700.
23. T. Richard, The Effect of Lignin on Biodegradability – Cornell Composting, https://compost.css.cornell.edu/calc/lignin.html (accessed 12 January 2023).
24. G. Singh, C. Chandoha-Lee, W. Zhang, S. Renneckar, P. J. Vikesland and A. Pruden, *Water Res.*, 2016, **104**, 137–146.
25. P. Khanjani, A. W. T. King, G. J. Partl, L.-S. Johansson, M. A. Kostiainen and R. H. A. Ras, *ACS Appl. Mater. Interfaces*, 2018, **10**, 11280–11288.
26. E. H. Baran and H. Y. Erbil, *Colloids Interfaces*, 2019, **3**, 43.
27. D. S. Aldana, E. D. Villa, M. De Dios Hernández, G. G. Sánchez, Q. R. Cruz, S. F. Gallardo, H. P. Castillo and L. B. Casarrubias, *Polymers*, 2014, **6**, 2386–2403.
28. Pretium Packaging, Plastics Comparison Chart.
29. E. Fortunati, D. Puglia, J. M. Kenny, M. Minhaz-Ul Haque and M. Pracella, *Polym. Degrad. Stab.*, 2013, **98**, 2742–2751.
30. S. Ghasemi, R. Behrooz, I. Ghasemi, R. S. Yassar and F. Long, *J. Thermoplast. Compos. Mater.*, 2018, **31**, 1090–1101.
31. P. Tyagi, L. A. Lucia, M. A. Hubbe and L. Pal, *Carbohydr. Polym.*, 2019, **206**, 281–288.
32. P. Sukyai, P. Anongjanya, N. Bunyahwuthakul, K. Kongsin, N. Harnkarnsujarit, U. Sukatta, R. Sothornvit and R. Chollakup, *Food Res. Int.*, 2018, **107**, 528–535.
33. M. B. Agustin, B. Ahmmad, E. R. P. De Leon, J. L. Buenaobra, J. R. Salazar and F. Hirose, *Polym. Compos.*, 2013, **34**, 1325–1332.
34. A. H. Tayeb, M. Tajvidi and D. Bousfield, *Molecules*, 2020, **25**, 1344.
35. A. Lewis, J. C. Waters, J. Stanton, J. Hess and D. Salas-De La Cruz, *Int. J. Mol. Sci.*, 2016, **17**, 9–12.
36. Y. Zhao, S. Xiao, J. Yue, D. Zheng and L. Cai, *Holzforschung*, 2020, **74**, 469–475.

37. N. Jeliazkova, M. Martinov, O. Tcheremenskaia, M. Networks, P. Rydberg, S. Avramova, N. Kochev, V. Jeliazkov and L. Iliev, Toxtree – Toxtree – Toxic Hazard Estimation by decision tree approach, https://toxtree.sourceforge.net/ (accessed 5 November 2022).
38. ECHA (European Chemicals Agency), Registration Dossier – 3,3,4,4,5,5,6,6,7,7,8,8,8-tridecafluorooctyl methacrylate, https://echa.europa.eu/registration-dossier/-/registered-dossier/14308 (accessed 9 January 2023).
39. M. Maras, C. Vanparys, F. Muylle, J. Robbens, U. Berger, J. L. Barber, R. Blust and W. De Coen, *Environ. Health Perspect.*, 2006, **114**, 100–105.
40. T. Serex, S. Anand, S. Munley, E. M. Donner, S. R. Frame, R. C. Buck and S. E. Loveless, *Toxicology*, 2014, **319**, 1–9.
41. ECHA (European Chemicals Agency), Registration Dossier – 3,3,4,4,5,5,6,6,7,7,8,8,8-tridecafluorooctan-1-ol, https://echa.europa.eu/registration-dossier/-/registered-dossier/5615 (accessed 9 January 2023).
42. TEDX, TEDX (The Endocrine Disruptione eXchange) List of Potential Endocrine Disruptors, https://endocrinedisruption.org/home/ (accessed 7 January 2023).
43. Pharos, 1H,1H,2H,2H-Perfluorooctan-1-ol, https://pharosproject.net/chemicals/2014372 (accessed 9 January 2023).
44. Pharos, 2-(*N*,*N*-diethylamino)ethyl methacrylate, https://pharosproject.net/chemicals/2004127 (accessed 9 January 2023).
45. ECHA (European Chemicals Agency), Registration Dossier – 2-(diethylamino)ethyl methacrylate, https://echa.europa.eu/lt/registration-dossier/-/registered-dossier/16027 (accessed 9 January 2023).
46. U.S. EPA (Environmental Protection Agency), CompTox Chemicals Dashboard, https://comptox.epa.gov/dashboard/.
47. DTSC, *Safer Consumer Products*, 2020, pp. 13–16.
48. OECD, *OECD Series on Risk Management*, 2020, 58, pp. 1–67.
49. M. A. Mohamed, W. N. W. Salleh, J. Jaafar, S. E. A. M. Asri and A. F. Ismail, *RSC Adv.*, 2015, **5**, 29842–29849.
50. J. P. Reddy and J. W. Rhim, *Carbohydr. Polym.*, 2014, **110**, 480–488.
51. D. Liu, T. Zhong, P. R. Chang, K. Li and Q. Wu, *Bioresour. Technol.*, 2010, **101**, 2529–2536.
52. C. Campano, R. Miranda, N. Merayo, C. Negro and A. Blanco, *Carbohydr. Polym.*, 2017, **173**, 489–496.
53. E. Gicquel, C. Martin, J. Garrido Yanez and J. Bras, *J. Mater. Sci.*, 2017, **52**, 3048–3061.
54. P. Tyagi, L. A. Lucia, M. A. Hubbe and L. Pal, *Carbohydr. Polym.*, 2019, **206**, 281–288.
55. J. A. Shatkin, K. J. Ong, J. D. Ede, T. H. Wegner and M. Goergen, *Tappi J.*, 2016, **15**(6), 425–437.
56. N. Stoudmann, M. Schmutz, C. Hirsch, B. Nowack and C. Som, *Nanotoxicology*, 2020, **14**, 1241–1257.
57. K. Kümmerer, J. Menz, T. Schubert and W. Thielemans, *Chemosphere*, 2011, **82**, 1387–1392.

58. IARC (International Agency for Research on Cancer), *Mists from Strong Inorganic Acids*, IARC, Lyon, FR, 2012.
59. ECHA (European Chemicals Agency), Registration Dossier – Sulphuric acid, https://echa.europa.eu/et/registration-dossier/-/registered-dossier/16122 (accessed 17 December 2020).
60. Pharos, Sulfuric acid, https://pharosproject.net/chemicals/2010413 (accessed 9 January 2023).
61. M. Roman, *Ind. Biotechnol.*, 2015, **11**, 25–33.
62. J. A. Shatkin and B. Kim, *Environ. Sci.: Nano*, 2015, **2**, 477–499.
63. S. Camarero-Espinosa, C. Endes, S. Mueller, A. Petri-Fink, B. Rothen-Rutishauser, C. Weder, M. J. D. Clift and E. J. Foster, *Fibers*, 2016, **4**, 21.
64. A. Tribot, G. Amer, M. Abdou Alio, H. de Baynast, C. Delattre, A. Pons, J.-D. Mathias, J.-M. Callois, C. Vial, P. Michaud and C.-G. Dussap, *Eur. Polym. J.*, 2019, **112**, 228–240.
65. N. D. Luong, N. T. T. Binh, L. D. Duong, D. O. Kim, D.-S. Kim, S. H. Lee, B. J. Kim, Y. S. Lee and J.-D. Nam, *Polym. Bull.*, 2012, **68**, 879–890.
66. Q. Wang, S. Xiao, S. Q. Shi and L. Cai, *Sci. Rep.*, 2018, **8**, 1–10.
67. A. H. Tayeb, M. Tajvidi and D. Bousfield, *Molecules*, 2020, **25**(6), 1344.
68. R. Datta, A. Kelkar, D. Baraniya, A. Molaei, A. Moulick, R. S. Meena and P. Formanek, *Sustainability*, 2017, **9**, 1163.
69. ECHA (European Chemicals Agency), Registration Dossier – 1H-Imidazolium, 3-ethyl-1-methyl-, chloride (1:1), https://echa.europa.eu/lt/registration-dossier/-/registered-dossier/16488 (accessed 9 January 2023).
70. National Center for Biotechnology Information, PubChem Compound Summary for CID 11321106, 1-Allyl-3-methylimidazolium chloride, https://pubchem.ncbi.nlm.nih.gov/compound/11321106 (accessed 9 January 2023).
71. O. US EPA, Safer Chemical Ingredients List, https://www.epa.gov/saferchoice/safer-ingredients (accessed 23 September 2022).
72. M. P. Vinardell and M. Mitjans, *Int. J. Mol. Sci.*, 2017, **18**(6), 1219.
73. G. Janusz, A. Pawlik, J. Sulej, U. Świderska-Burek, A. Jarosz-Wilkołazka and A. Paszczyński, *FEMS Microbiol. Rev.*, 2017, **41**, 941–962.
74. C. M. Ewulonu, X. Liu, M. Wu and H. Yong, Lignin-containing cellulose nanomaterials: A promising new nanomaterial for numerous applications, *J. Bioresour. Bioprod.*, 2019, **4**, 3–10.
75. W. Yang, G. Qi, J. M. Kenny, D. Puglia and P. Ma, *Polymers*, 2020, **12**, 1364.
76. Siyavula Education, *Paper making process*, 2014.
77. A. H. Bhat, Y. K. Dasan, I. Khan, H. Soleimani and A. Usmani, in *Cellulose-Reinforced Nanofibre Composites: Production, Properties and Applications*, Elsevier Inc., 2017, pp. 215–240.
78. S. Bellamkonda, *Curry Chickpea Bowl @sweetgreen*, 2017, https://www.flickr.com/photos/drbeachvacation/32574005580/.
79. A. Salles, *Burger and fries*, 2015, https://www.flickr.com/photos/132313505@N07/20554610314/.
80. W. Jeffrey, *Cherry tomatoes on pasta*, 2022, https://www.flickr.com/photos/jeffreyww/14530170849/.

CHAPTER 4

PFAS-free Moisture Barriers in Household Product Packaging

A. GUAN*[a] AND K. G. HAZARD[b]

[a] Department of Materials Science and Engineering, University of California, Berkeley, CA, 94720, USA; [b] University of California, Berkeley, School of Public Health, Berkeley, CA, 94704, USA
*Email: amandaguan@berkeley.edu

4.1 Challenge Overview

This case study highlights a collaboration between University of California (UC) Berkeley students and household product manufacturer, Method, to address the challenge of designing a safer alternative to moisture barriers that use perfluoroalkyl and polyfluoroalkyl substances (PFASs) for a selection of the company's product packaging. The objective was to research and analyze alternatives to the PFASs and other petrochemicals currently used as moisture barriers applied on paper-based packaging. Currently, the company relies on recycled plastic bottles and pouches to house their naturally-derived, biodegradable, and vegan formulations. However, in an effort to continue pioneering greener solutions, Method has expressed interest in moving towards cellulose-based paper packaging. While paper offers sufficient rigidity and can be easily recycled or composted at the end of its life, it has poor barrier properties, which have traditionally been remedied by the addition of PFASs, low density polyethylene (LDPE), and similar materials into the packaging. Alternatives derived from environmentally sustainable resources and compatible with municipal recycling or composting systems were prioritized.

Green Chemistry Series No. 81
Toward a PFAS-free Future: Safer Alternatives to Forever Chemicals
Edited by Simona A. Bălan, Thomas A. Bruton and Kimberly G. Hazard

Published by the Royal Society of Chemistry, www.rsc.org

This project comprised a semester-long investigation of alternatives to PFAS-based product packaging as part of the UC Berkeley graduate course, Greener Solutions, from August to December 2020. Course instructors (Dr Megan Schwarzman, Mr Thomas McKeag, Dr William Hart-Cooper, and Ms Kimberly Hazard) facilitated and guided the partnership between Kaj Johnson, Senior Director of Research and Development at Method, and the student team, consisted of Constancia Dominguez (Environmental Health Sciences), Amanda Guan (Materials Science and Engineering), Grayson Hamaker-Teals (Chemical Biology), and Josue Ruiz (Epidemiology & Biostatistics).

Aligning their search with industry needs, the student team selected the three most common Method product packaging types for investigation: dry powders (*e.g.*, laundry powder), concentrated liquids (*e.g.*, laundry detergent), and dilute, high water content solutions (*e.g.*, dish soap).

Success criteria for a moisture barrier material or treatment were defined as follows:

1. Comparatively adequate moisture barrier (low water vapor permeability and high water contact angle, approximating LDPE performance).
2. Low environmental hazards (renewably sourced, biodegradable, low aquatic toxicity, low persistence, and for Method specifically, vegan compliance).
3. Reduced health impacts (non-irritating, non-carcinogenic or mutagenic, low reproductive/developmental toxicity, and no endocrine disruption).

A viable alternative should function as a barrier throughout the product's expected shelf life, rather than matching the same (potentially over-engineered) physical performance capabilities as the current industry standards. The solutions were evaluated by directly comparing select "bad actors" with the proposed alternatives in terms of both technical performance and chemical hazard, as summarized in the tables that follow. Three strategies were proposed, designed to build off one another, starting with a biopolymer base and later blending chemical and physical additives to further improve barrier and mechanical properties as needed.

Method products were categorized by dilution and water content, which directly indicated each product's moisture barrier requirement (as shown in Figure 4.1). Dry powders, such as powdered laundry detergents and other solid formulations, had the lowest moisture barrier threshold. Although these solid powders are unlikely to leak out of a package in the same manner as a liquid formulation, they may still draw in moisture and cake, thus ruining the product before its use and deteriorating the packaging. The next category in the spectrum of increasing dilution consists of concentrated liquids such as liquid laundry detergents. This type of formulated products requires a medium level of moisture barrier strength to keep liquids contained. The highest barrier demand is required for high water content

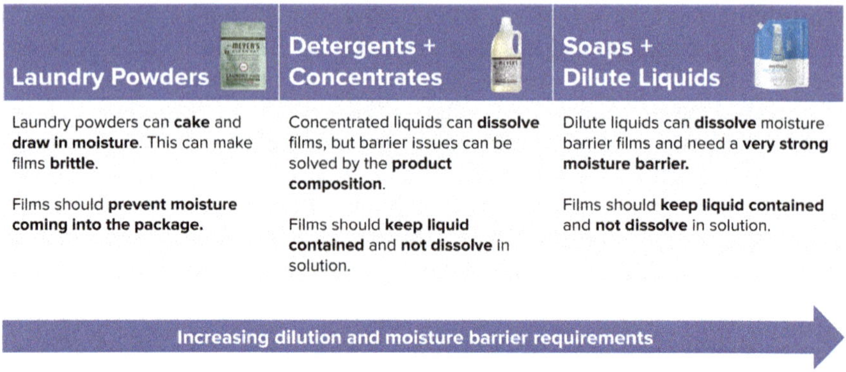

Figure 4.1 Moisture barrier spectrum for three different types of method products. Films for paper packaging should aim to contain the product without causing any damage to the primary packaging.

products, such as dilute soaps. In this case, films should keep the liquid contained and ideally not dissolve in solution. Although each of the three strategies identified by the team applies best to one of these specific product categories, the film strategy designed to protect the highest liquid content would also be suitable for the products with a lower moisture barrier threshold.

4.2 Importance, Innovation, and Impact

Packaging remains the largest generator of plastic waste.[1] Thus, employing greener alternatives to traditional household product packaging can reduce the amount of single-use plastic waste generated and minimize harmful exposures to PFASs and other toxicants.

The fiber-based packaging market is growing globally, with paper packaging emerging as a preferred alternative to petroleum-based plastic packaging.[2] Cellulose-based packaging requires minimal energy to recycle, maintains structural integrity, is low cost, and biodegrades. However, because of its fibrous and hydrophilic nature, paper packaging cannot adequately address moisture barrier needs for a variety of applications, such as formulated products. As a result, the packaging industry still relies heavily on polyolefins, such as polyethylene and propylene, as well as on PFASs as treatments and coatings for improved barrier properties.

Although polyolefins are recyclable on their own, once added to a separate packaging material they render the entire packaging system non-biodegradable, non-recyclable, and ultimately persistent in the environment. The impacts of polyolefins on human health and the environment outweigh the advantages of these materials' moisture barrier function. The adoption of PFASs as the default moisture barrier has perhaps led to slowing of innovation across consumer product industries, including household and food packaging. In many applications, select PFASs over-perform as moisture barriers and adversely impact the health of workers, consumers, and the environment. The "one size fits all" application of these substances could be replaced with more selective formulations tuned to the specific technical requirements of packaging for different types of products, *e.g.*, dry powders, highly concentrated liquids, and high water content liquids. The alternatives highlighted in this chapter offer a range of possible safer solutions to three different types of packaging systems, as well as an approach that gives equal weightage to environmental, health, and technical performance, all without the use of polyolefins or PFASs.

4.3 Solutions

A three-prong approach is proposed to address a range of barrier property needs: biopolymer films derived from natural sources (for dry powders), cross-linking chemical additives to improve biopolymer film properties (for concentrated liquid solutions), and nanofiller physical additives (for high water content liquids). These represent an opportunity map for addressing a range of moisture barrier needs from which industry can select any or all options as suited.

4.3.1 Biopolymer-based Films

Biopolymer-based films were initially selected for their biodegradability and low toxicity, and they serve as the foundation upon which the other strategies are built. Biopolymers, such as those selected for investigation in this project (chitosan, gelatin, and pectin), can be sourced from plants, are often safe for human consumption, and encourage a circular economy by recovering and repurposing waste products. Their polymeric nature also presents the opportunity to blend chemical and physical additives (as described in Sections 4.3.2 and 4.3.3, respectively) to improve both mechanical and barrier properties.

4.3.1.1 Technical Performance

To establish a comprehensive foundation for a sustainable moisture barrier solution, the first alternative was analyzed through three biopolymer candidates – chitin/chitosan, pectin, and gelatin – which were chosen because of their availability in nature and previous indications in the literature of their potential as biodegradable films. Their success was evaluated by comparing their technical properties to those of industry leading "bad actors": polytetrafluoroethylene (PTFE), a popular PFAS used in non-stick applications, as well as polyethylene and polypropylene. Polyethylene and polypropylene have excellent sealing properties for paper, and are the most common form of synthetic polymers used for film coatings.[3] Data for this comparative technical assessment were sourced from a variety of peer-reviewed literature and scientific databases.

Chitosan, the deacetylated form of chitin, has a water vapor permeability (WVP) several orders of magnitude higher (worse) than that of leading industry coatings but demonstrates a comparably high water contact angle.[4,5] As for mechanical properties, chitosan has a slightly lower tensile strength and elasticity than bad actors.[6,7] Pectin, the next biopolymer alternative evaluated, demonstrated a WVP several orders of magnitude higher than that of the industrial bad actors, similar to chitosan.[8] Pectin also has a water contact angle below 90 degrees, indicating lack of required hydrophobicity to be an appropriate moisture barrier. Mechanically, pectin also fell short, with a medium efficiency tensile strength and low elasticity. The final alternative, gelatin, lacked specific quantitative data for its moisture barrier properties but demonstrated high tensile strength, exceeding even current industry solutions; conversely, gelatin has very poor elasticity compared to the bad actors, which may prove troubling in product packaging.[9] These summarized results can be found, along with more specific quantitative data, within Figure 4.2. The studied biopolymer alternatives demonstrated inferior barrier and mechanical properties to PTFE and polyolefins, indicating a need for additives to improve their performance.

4.3.1.2 *Hazard Assessment*

Although the biopolymer alternatives had some technical shortcomings, they do have highly favorable hazard profiles. Perfluorooctanoic acid (PFOA) (a PFAS traditionally used as a processing aid in the manufacture of polymeric PFASs such as PTFE), polyethylene, and polypropylene were used as baseline chemicals to evaluate hazards. It should be noted that PTFE was used as the PFAS baseline for technical comparison, while PFOA is used for hazard assessment. This is because PTFE, as a thin film used in products, is inert with minimal hazards, whereas PFOA is a residual or impurity associated with PTFE with higher potential for exposure, making it pertinent to analyze its effects in human and environmental systems.

PFOA has been shown to have a multitude of health concerns due to its high resistance to environmental degradation, high incidence of bioaccumulation, long-term toxic effects in aquatic organisms and potential links between PFOA exposure and cancer in both humans and laboratory animals.[9,16–18] While polyethylene and polypropylene are effective moisture barriers, such polyolefin coatings cause an entire packaging system to lose its biodegradability (if desired for composting) and recyclability (as many municipalities do not accept #4 plastic or blended materials for curbside recycling).

Due to its biodegradability, lack of carcinogenicity, and more environmentally sustainable nature, chitosan emerges as a greener alternative compared to PFASs, polyethylene, and polypropylene.[19,20] Chitosan is classified as potentially hazardous to aquatic environments under the Global Harmonized System of Classification and Labeling of Chemicals (GHS) due to toxicity studies in which oxygen interference and physiological disorders

Strategy 1: Technical Performance Comparison	Barrier Properties		Mechanical Properties	
	Water Vapor Permeability (WVP) (g/m*day*atm)	Water Contact Angle	Tensile Strength (MPa)	Total Elongation at Break
Baseline: PTFE	0.00788 (H)[10]	106.94° (H)[11]	10.0 - 45.0 (H)[12]	40.0 - 650% (H)[12]
Baseline: Polyethylene	LDPE: 0.008 HDPE: 0.002 (H)[13]	LDPE: 91° HDPE: 93° (H)[5]	LDPE: 13.2 HDPE: 13.9 (H)[14]	LDPE: 456% HDPE: 334% (H)[14]
Baseline: Polypropylene	0.0575 (H)[10]	97° (H)[5]	18 – 22 (H)[15]	50 - 145% (H)[15]
Alternative: Chitin / Chitosan	.315 (L)[4]	82-104° (H)[5]	Neat: 37.7 In 2% Solution: 6.99 (M)[6,7]	Neat: 49.5% In 2% solution: 72.70% (M)[6,7]
Alternative: Pectin	0.135 (L)[8]	62.1° (M)[8]	7.10 (M)[8]	7.17% (L)[8]
Alternative: Gelatin	Non-ideal mechanical properties and water vapor barrier (M)[13]		70 (H)[9]	1.5% (L)[9]

Key	Efficiency	High (H): Similar or exceeding properties of current use substances	Moderate (M): Slightly worse	Low (L): Much worse, by order of magnitude

Figure 4.2 Technical performance comparison for three baseline substances (PTFE,[10–12] polyethylene,[5,13,14] and polypropylene[5,10,15]) and three proposed alternative biopolymers (chitin/chitosan,[4–7] pectin,[8] and gelatin[9,13]). LDPE = low density polyethylene and HDPE = high density polyethylene.

in various aquatic animals were noted.[21,22] However, the German Federal Environment Agency (FEA) lists chitosan as non-hazardous to waters (note that this is not strictly speaking a list of aquatic toxicants, but rather of chemicals that fit a variety of chemical hazard profiles that should not enter the waters, whether due to aquatic toxicity or toxicity to humans who will use the waters).[23] While contradictory data make it difficult to assess the true aquatic toxicity of chitosan, the student team deferred to the authoritative GHS classification.

Next, pectin, a natural polysaccharide derived from citrus peels and apple pomace, was found to be biodegradable and not persistent in the environment. Studies from the European Food Safety Authority (EFSA) Panel on Food Additives and Nutrient Sources reveal that the biopolymer pectin has no chronic toxicity effects.[24]

Finally, gelatin, also derived from organic material, is classified by the Australian National Industrial Chemicals Notification and Assessment Scheme as a low concern to human health and environment on the basis of the tier 1 assessment (high throughput assessments).[25] The overall comparative hazard assessment is summarized in Figure 4.3, featuring both current industry solutions and biopolymer alternatives. All three alternatives showed a dramatic reduction in hazards in comparison to PFASs and polyolefins, while also being readily biodegradable and naturally sourced.

4.3.1.3 Feasibility

When projected onto the Method product line, biopolymer film alternatives, while posing minimal overall hazards, lack the barrier and mechanical properties necessary to form a sufficient film for liquid products, such as detergents and soaps. However, these films would still be a minimum viable solution suitable for dry laundry powders, as they could prevent moisture from being drawn in.

4.3.2 Addition of Cross-linking Reagents

To mitigate the technical concerns that arise with biopolymer films alone, chemical cross-linking reagents could be added to improve mechanical and barrier properties of the packaging. Covalent bonding of different biopolymer structures with chemical cross-linking reagents can strengthen dry hydrogels and create a polyion-complex due to the reagents' charged interactions.[18] The resulting network structure of a cross-linked polymer also increases the diffusion path length for moisture to cross a barrier, thus decreasing its water vapor permeability. For this alternative assessment, glutaraldehyde was selected as the chemical baseline for technical and hazard assessment because of its prevalence as a cross-linker in current industrial film formulations. Alternatives were evaluated based on how efficient they were in improving the mechanical and barrier properties of the biopolymers identified in Section 4.3.1.

Strategy 1: Hazard Assessment	**Persistence**	**Bioaccumulation**	**Sensitivity / Irritation** *(Eye, Skin, Resp.)*	**Toxicity** *(Developmental & Reproductive, Systemic, Neuro.)*	**Aquatic Toxicity**	**Carcinogen / Mutagen**	**Endocrine Activity**
Baseline: PFOA	H[26]	H[27]	M[18]	H[16,17]	H[18]	H[18]	H[18]
Baseline: Polyethylene	H[28]	L[29]	M[30]	DG	DG	H-L[31]	DG
Baseline: Polypropylene	H[28]	DG	M[32]	DG	DG	H-L[33]	DG
Alternative: Chitin / Chitosan	L[34]	L[34]	M[20]	L[20]	M[21, 22]	L[20]	DG
Alternative: Pectin	L[35]	L[35]	M[36]	L[24]	L[35]	L[35]	DG
Alternative: Gelatin	L[37]	L[37]	DG	DG	L[37]	L[37]	DG

Key	Hazard	High (H)	Moderate (M)	Low (L)	Data Gap (DG)

Figure 4.3 Hazard assessment comparison for three baseline substances (PFOA,[16–18,26,27] polyethylene,[28–31] and polypropylene[28,32,33]) and three proposed alternative biopolymers (chitin/chitosan,[20–22,34] pectin,[24,35,36] and gelatin[37]).

4.3.2.1 Technical Performance

The first cross-linking alternative investigated is genipin, a derivative of the gardenia fruit. It was chosen because of its success in the medical industry as a tissue fixative and, according to more recent studies, for its application in moisture barrier films. Gelatin-pectin blended films cross-linked with genipin showed significantly lower degrees of swelling in water.[38] Chitosan/polyethylene oxide films cross-linked with genipin showed an increase in tensile strength and a slight, although less significant, increase in elongation at break.[39] While research surrounding genipin as a cross-linking reagent is limited, the evidence suggests that genipin-cross-linked films outperform biopolymer films without additives in both mechanical and barrier properties.

The second cross-linking reagent investigated is ferulic acid, which is also naturally derived. When added to a gelatin film, ferulic acid increased tensile strength, however, offered no recognizable improvements in WVP.[40] In chitosan film applications, ferulic acid actually lowered efficiency for barrier and elongation, while increasing tensile strength.[41] Due to significant data gaps and lack of research, no conclusive assessment could be reached for ferulic acid's impact on pectin films. Based on technical performance, ferulic acid is likely not an appropriate cross-linking solution to replace leading industry cross-linking agents. Figure 4.4 shows detailed technical comparisons between glutaraldehyde and the two alternatives discussed.

4.3.2.2 Hazard Assessment

Glutaraldehyde was used again as an industry baseline to conduct a comparative human and environmental health assessment, with results available in Figure 4.5. Studies have identified glutaraldehyde as cytotoxic and neurotoxic, as well as a chronic toxicant for aquatic organisms.[48–50] A study directly comparing the cross-linking capabilities of genipin and glutaraldehyde for chitosan-based films concluded that genipin was approximately 10 000 times less cytotoxic than glutaraldehyde.[39] There is a general data gap for bioaccumulation, carcinogenicity/mutagenicity, as well as its potential as an endocrine disruptor; however from the information available, genipin proved to be significantly less hazardous than glutaraldehyde. In the case of ferulic acid, no adverse effects were observed in data on aquatic organisms or humans reported by authoritative agencies. As a natural cross-linking alternative, ferulic acid was also found to be less environmentally and biologically persistent, in addition to having a minimal hazard potential for endocrine disruption, aquatic and developmental toxicity, and carcinogenicity and mutagenicity. Overall, both the alternatives demonstrated lower environmental persistence and toxicity compared to glutaraldehyde.

Technical Performance Comparison (Strategy 2)		Barrier Properties		Mechanical Properties	
Cross-linker	**Biopolymer**	**Water Vapor Permeability (WVP) (g*mm/kPa *m2* hr)**	**Water Contact Angle**	**Tensile Strength (MPa)**	**Total Elongation at Break**
Baseline: **Glutaraldehyde (Glu)**	Pectin	DG	DG	Increase from 11.1 to 21.6 with 0.3 wt% addition of Glu in gelatin-pectin blend (H)[38]	Increase from 151.57 to 159.1 with 0.3 wt% addition of Glu in gelatin-pectin blend (H)[38]
	Gelatin	Decrease from 1.8 to 0.8 with 0.75% addition of Glu (H)[42]	DG	Increase from ~1.2 to ~3.2 with 2.5 wt% addition of Glu (H)[43]	Decrease from ~35 (0.05 wt% Glu) to ~12 (2.5 wt% Glu) (L)[43]
	Chitosan	Increase from ~111 to ~115 with 33 wt% addition of Glu (L)[44]	Increase from 90 to 98 with 33 wt% addition of Glu (H)[44]	Increase from 32.9 to 44.1 with 5 wt% addition of Glu (H)[45]	Decrease from 38.6 to 30.3 with 5 wt% addition of Glu (L)[45]
Alternative: **Genipin (Gen)**	Pectin	Lower water sensitivity in gelatin-pectin films (H)[46]	DG	DG	DG
	Gelatin		DG	Little to no effect, stays at 1.0, with 2 wt% addition of Gen (M)[47]	Decrease from 211 to 13 with 2 wt% addition of Gen (L)[47]
	Chitosan	Little to no effect, stays at ~115, with 33 wt% addition of Gen (M)[44]	Increase from ~90 to ~96 with 33 wt% addition of Gen (H)[44]	Increase from ~39 to ~50 with 0.5 wt% addition of Gen in dry film (H)[39]	Little to no effect, from ~9 to ~10, with 0.5 wt% of Gen in dry film (M)[39]
Alternative: Ferulic Acid (FA)	Pectin	DG	DG	DG	DG
	Gelatin	Little to no effect, from 0.00208 to 0.00201, with 40 wt% addition FA (M)[40]	DG	Increase from 85 to 95.5 with 40 wt% addition of FA (H)[40]	Decrease from 4.5 to 3 with 40 wt% addition of FA (L)[40]
	Chitosan	Increase from 0.049 to 0.075 with 0.08 wt% addition of FA-coupled chitosan (L)[41]	DG	Increase from 15.99 to 20.71 with 0.08 wt% addition of FA-coupled chitosan (H)[41]	Decrease from 21.39 to 9.34 with 0.08 wt% addition of FA-coupled chitosan (L)[41]

Key	Efficiency	High (H): Improves properties of base film	Moderate (M): Little to no change to properties of base film	Low (L): Worsens properties of base film	Data Gap (DG)

Figure 4.4 Technical performance comparison of the current baseline cross-linker (glutaraldehyde[38,42–45]) and two alternative cross-linkers (genipin[39,44,46,47] and ferulic acid[40,41]) paired with three proposed alternative biopolymers.

Strategy 2: Hazard Assessment	Persistence	Bioaccumulation	Sensitivity / Irritation *(Eye, Skin, Respiratory)*	Toxicity *(Developmental & Reproductive, Systemic, Neuro.)*	Aquatic Toxicity	Carcinogen / Mutagen	Endocrine Activity
Baseline: **Glutaraldehyde**	M[51]	L[52]	M[53]	H[49]	H[50]	DG	H[54]
Alternative: **Genipin**	L[55]	DG	DG	L[55,39]	DG	DG	DG
Alternative: **Ferulic Acid**	L[41]	DG	M[56]	L[56]	L[56]	L[56]	L[56]

Key	Hazard	High (H)	Moderate (M)	Low (L)	Data Gap (DG)

Figure 4.5 Comparative hazard assessment for current baseline cross-linker (glutaraldehyde[49–54]) and two proposed alternative cross-linking reagents (genipin[39,55] and ferulic acid[41,56]).

4.3.2.3 Feasibility

While some data gaps exist for a few hazard endpoints and technical properties, the available data demonstrate that the presented alternatives are indeed safer than glutaraldehyde. Both alternatives performed less efficiently than glutaraldehyde, with ferulic acid demonstrating particularly low performance. Still, especially in the case of genipin, using a safer cross-linker enhanced the properties of a biopolymer matrix, making this alternative more appropriate for applications with moderate moisture barrier needs, such as concentrated liquids.

4.3.3 Biopolymer Films with Physical Additives

4.3.3.1 Technical Performance

The final alternative centers around the incorporation of nanofillers (*i.e.*, natural clays and fibers) into a biopolymer matrix to form nanocomposites, which studies have shown to enhance moisture barrier properties while still maintaining environmental degradability. Figure 4.6 demonstrates how silicate layers (*i.e.*, clays) can be blended with polymers to create different types of microcomposite and nanocomposite films.[57] Exfoliated nanocomposite films generally result in the highest improvement in barrier properties, as they create the longest path for particles, such as water, to travel through.[58]

One nanofiller of interest is cellulose nanocrystals (CNCs), which are synthesized from cellulose fibers into nanocrystals *via* hydrolysis. This results in a biomaterial with nanoscale dimensions, unique morphology, high surface area, low density, biodegradability, renewability, and high mechanical strength.[59] CNCs are processed with acid treatment, to remove the para-crystalline domains regularly distributed along the microfibers,

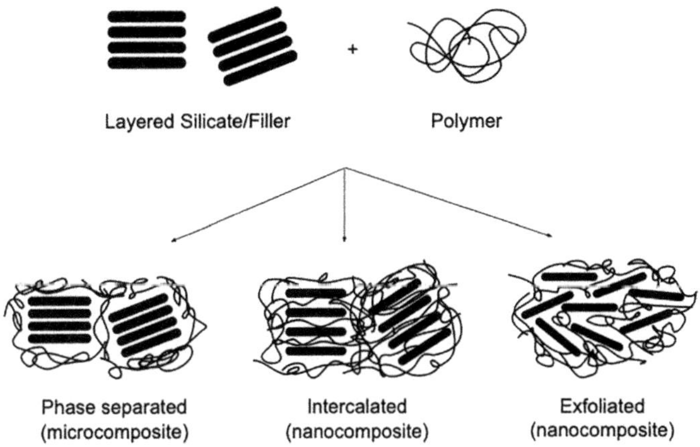

Figure 4.6 Scheme of three types of composite structures: phase separated microcomposites, intercalated nanocomposites, and exfoliated nanocomposites. Reproduced from ref. 58, https://doi.org/10.1080/23311916.2021.1991229, under the terms of the CC BY 4.0 license https://creativecommons.org/licenses/by/4.0/.

and then physically added into various polymer matrices to form polymer nanocomposites.[60]

Montmorillonite, a natural clay, was selected as another sustainable nanofiller for its desirable qualities, such as its ability to strengthen biopolymer films and reinforce material in a naturally abundant, inexpensive, and chemically and thermally stable manner.[28] Figure 4.7 compares the technical performance of the two proposed alternative physical additives in each of the three alternative biopolymers. Montmorillonite generally shows a positive trend of positive efficiency for all performance metrics, aside from total elongation at break.[29–31] Similarly, micro- and nano-fibrillated cellulose (NFC/MFC) or CNCs provided high efficiency for WVP and tensile strength.[26,32] Overall, addition of both montmorillonite and cellulose nanocrystals to biopolymer films resulted in similar improvements in barrier properties and tensile strength.

4.3.3.2 Hazard Assessment

The physical additives, montmorillonite and CNCs, each demonstrated health and environmental improvements over current industrial films, as shown in Figure 4.8. Montmorillonite was found to have low overall hazards[69] and has been proposed for therapeutic drug delivery due to its nontoxic and detoxifying properties.[70] CNCs were also found to be of low toxicity and do not bioaccumulate nor persist.[71] The main concern pertaining to the human health performance of clay and fiber nanofillers is the occupational risk of exposure during the manufacturing stage. The Association of Occupational and Environmental Clinics (AOEC) classifies CNCs as an asthmagen,[71] and occupational montmorillonite exposure may be associated with the development of pneumoconiosis.[72] Occupational risks to respirable

Strategy 3: Technical Performance Comparison		Barrier Properties		Mechanical Properties	
Physical additive	**Biopolymer**	**Water Vapor Permeability (WVP) (g*mm/kPa *m2* hr)**	**Water Contact Angle**	**Tensile Strength (MPa)**	**Total Elongation at Break**
Alternative: Montmorillonite (MMT)	Pectin	Decrease from 2.52 to 1.51 with 8 wt% addition of MMT (H)[61]	DG	Increase from 2.4 to 3.5 with 8 wt% addition of MMT (H)[61]	Decrease from 6.6 to 5.4 with 8 wt% addition of MMT (L)[61]
	Gelatin	Decrease from 6.2×10^{-13} to 1.8×10^{-13} with 15 wt% addition of MMT (H)[62]	DG	Increase from 10 to 38 with 0.5 wt% addition of MMT (H)[63]	Decrease from 38 to 30 with 0.5 wt% addition of MMT (L)[63]
	Chitosan	Decrease from 2.6×10^{-7} to 1.6×10^{-7} with 5 wt% addition of MMT (H)[64]	DG	Increase from 61 to 69 in 5 wt% addition of MMT(H)[64]	Decrease from 3.8 to 3.0 % in 5% wt% addition of MMT (L)[64]
Alternative: Cellulose Nanocrystals (CNC)	Pectin	Decrease from 4.6×10 -7 to 3.3×10−7 with 5 wt% addition of the CNC (H)[65]	DG	Increase from 7.1 to 13.2 with 5 wt% addition of CNC (H)[65]	Increase from 20 to 30% with 5 wt% addition of CNC (H)[65]
	Gelatin	Dec. from 2.2×10^{-7} to 1.6 x × 10^{-7} with 4 wt% addition of CNC (H)[59]	DG	Increase from 83 to 108 with 4 wt% addition of CNC (H)[59]	Dec. from 38 to 23% with 4 wt% addition of CNC (L)[59]
	Chitosan	Decrease by 45% with 3 wt% addition of CNC (H)[66]	Little to no effect (from ~85 to ~84) with the addition of 15 wt% CNC (M)[67]	Increase from 79 to 98 with 10 wt% addition of CNC (H)[68]	DG

Key	Efficiency	High (H): Improves properties of base film	Moderate (M): Little to no change to properties of base film	Low (L): Worsens properties of base film

Figure 4.7 Technical performance comparison of two proposed alternative physical additives (montmorillonite[61–64] and cellulose nanocrystals[59,65–68]).

Strategy 3: Hazard Assessment	Persistence	Bioaccumulation	Sensitivity / Irritation *(Eye, Skin, Respiratory)*	Toxicity *(Developmental & Reproductive, Systemic, Neuro.)*	Aquatic Toxicity	Carcinogen / Mutagen	Endocrine Activity
Alternative: Montmorillonite (MMT)	DG	DG	M[72]	DG	L[69]	L[69]	DG
Alternative: Cellulose Nanocrystals (CNC)	L[71]	L[71]	H[36]	L[67]	L[73]	L[71]	DG

Key	Hazard	High (H)	Moderate (M)	Low (L)	Data Gap (DG)

Figure 4.8 Comparative hazard assessment for two proposed alternative physical additives (montmorillonite[69,72] and cellulose nanocrystals[36,67,71,73]).

nanofillers may be mitigated by appropriate personal protective equipment and engineering controls. Overall, both montmorillonite and CNCs showed a decrease in the hazard level compared to the currently used films.

4.3.3.3 *Feasibility*

Nanocomposites of biopolymer films, as exemplified by montmorillonite and CNCs, showed great potential for an enhanced water vapor barrier, while still maintaining the biodegradability of the film. This alternative showed even greater mechanical and barrier improvements than previous alternatives, meaning it would rank highest in the moisture barrier spectrum, and would be the best option of the three discussed to replace polyolefin films in dilute liquid packaging. Ultimately, additional experimental research is required to identify the potential success nanocomposites could display as a moisture barrier in product packaging. Neither individual additive matched up to leading industry polyolefin solutions, but considering that such solutions are often overengineered, this alternative could still meet industry standard performance criteria.

4.4 Final Recommendations

The journey from the status quo toward a PFAS-free future is a steep climb. The following considerations acknowledge the complexity of such an endeavor. First, renewable sources for biopolymer films may initially pose an additional cost to manufacturers, but they can assist in reducing long term costs. For example, waste products can be collected for their high chitin content and used as the basis for the biopolymer films. Second, filling out remaining data gaps for both health hazard assessment and technical performance will ensure that the performance of the alternatives properly matches the current methods in terms of packaging stability, without contributing to human or environmental health risks. The project highlighted here was a literature-exclusive study that was limited in scope. Next steps would include testing a combination of biopolymers and additives, not necessarily limited to those identified in this report, incorporated within films to develop packaging suitable for a diverse array of dry powders and formulated liquids.

More immediately, it was recommended that the project partner identify packaging solutions that may "overperform" in terms of their hydrophobicity and oleophobicity. For instance, could these products integrate a biopolymer film that performs at a lower level compared to polyolefins without sacrificing their market viability? Would sustainability-inclined consumers agree to reduced barrier properties if they were informed about the environmental benefits of the packaging? Answering these questions would require some degree of market research and consumer feedback.

Based on the analysis in this report, a chemically cross-linked biopolymer film could be a viable alternative for detergents, which have a moderate demand for moisture barrier. Experimenting with different forms of cross-linking such as physical, chemical, and enzymatic methods may prove useful in identifying the ideal biopolymer combination and cross-linking reagent. With sufficient iteration, biopolymer films with montmorillonite or CNCs may be suitable to contain method products such as the dish soap refills, which have the highest level of moisture barrier requirements.

Finally, while not within the scope of this investigation, there are sustainability strategies that can reduce carbon emissions by circumventing the need for single-use packaging altogether, such as concentrated, solid alternatives that consumers can dilute at home. An estimated 20% of global plastic packaging by weight could be replaced with reusable packaging if only active ingredients were shipped, resulting in roughly 97% less water transported and 87% fewer trucks on the road.[74] Additionally, the barrier needs of concentrated, solid alternatives can easily be met with the above-proposed moisture barriers. However, some obstacles for this paradigm shift include access to clean water and company product development infrastructure and timelines. The search for greener, PFAS-free alternatives to plastic household packaging remains a worthy and necessary pursuit that not only requires robust research and experimentation, but also ingenuity and creativity.

References

1. A. Hogan and A. Steinbach, A Polymer Problem, https://www.law.georgetown.edu/environmental-law-review/blog/a-polymer-problem-how-plastic-production-and-consumption-is-polluting-our-oceans/ (accessed 27 July 2022).
2. Transparency Market Research, Fiber Based Packaging Market Bolsters with Rise in Demand for Environment-friendly Products, States TMR Study,

https://www.globenewswire.com/news-release/2022/03/08/2398940/0/en/Fiber-Based-Packaging-Market-Bolsters-with-Rise-in-Demand-for-Environment-friendly-Products-States-TMR-Study.html (accessed 27 July 2022).

3. K. Khwaldia, E. Arab-Tehrany and S. Desobry, *Compr. Rev. Food Sci. Food Saf.*, 2010, **9**, 82–91.
4. B. L. Butler, P. J. Vergano, R. F. Testin, J. M. Bunn and J. L. Wiles, *J. Food Sci.*, 1996, **61**, 953–956.
5. N. Encinas, M. Pantoja, J. Abenojar and M. A. Martínez, *J. Adhes. Sci. Technol.*, 2010, **24**, 1869–1883.
6. S.-F. Wang, L. Shen, W.-D. Zhang and Y.-J. Tong, *Biomacromolecules*, 2005, **6**, 3067–3072.
7. K. Gadgey and G. Sharma, *Int. J. Adv. Res. Eng. Technol.*, 2017, **8**, 93–102.
8. T. M. Ngo, T. H. Nguyen, T. M. Dang, T. X. Tran and P. Rachtanapun, *Int. J. Mol. Sci.*, 2020, **21**(6), 2224.
9. M. Echegaray, G. Mondragon, L. Martin, A. González, C. Peña-Rodriguez and A. Arbelaiz, *J. Renewable Mater.*, 2016, **4**(3), 206.
10. P. E. Keller and R. T. Kouzes, *Water Vapor Permeation in Plastics*, Pacific Northwest National Lab. (PNNL), Richland, WA (United States), 2017.
11. J. Włoch, A. P. Terzyk, M. Wiśniewski and P. Kowalczyk, *Langmuir*, 2018, **34**, 4526–4534.
12. Properties of Molded PTFE – Trebor, https://treborintl.com/properties-for-molded-ptfe/ (accessed 18 October 2022).
13. S. C. Shit and P. M. Shah, *J. Polym.*, 2014, **2014**, 1–13.
14. Online Materials Information Resource – MatWeb, https://www.matweb.com/index.aspx (accessed 15 October 2022).
15. polymerdatabase.com, Polymer Properties Database, https://polymerdatabase.com/home.html (accessed 3 November 2020).
16. F. M. Hekster, P. De Voogt, A. M. C. M. Pijnenburg and R. W. P. M. Laane, *Perfluoroalkylated substances: Aquatic environmental assessment*, Rijkswaterstaat, RIKZ, 2002.
17. V. M. Vieira, K. Hoffman, H.-M. Shin, J. M. Weinberg, T. F. Webster and T. Fletcher, *Environ. Health Perspect.*, 2013, **121**, 318–323.
18. Agency for Toxic Substances and Disease Registry (ATSDR), *Toxicological Profile for Perfluoroalkyls: Health Effects*, 2021.
19. M. N. V. R. Kumar, *React. Funct. Polym.*, 2000, **46**, 1–27.
20. National Toxicology Program, *NTP Technical Report on the Toxicity Study of Chitosan (CASRN 9012-76-4) Administered in Feed to Sprague Dawley [Crl:CD(SD)] Rats*, National Toxicology Program, Research Triangle Park (NC), 2017.
21. G. Bullock, V. Blazer, S. Tsukuda and S. Summerfelt, *Aquaculture*, 2000, **185**, 273–280.
22. Y.-L. Hu, W. Qi, F. Han, J.-Z. Shao and J.-Q. Gao, *Int. J. Nanomed.*, 2011, **6**, 3351–3359.

23. Umweltbundesamt, Search Details – Rigoletto, http://webrigoletto.uba.de/Rigoletto/Home/SearchDetail/2986 (accessed 5 December 2022).
24. A. Mortensen, F. Aguilar, R. Crebelli, A. Di Domenico, B. Dusemund, M. J. Frutos, P. Galtier, D. Gott, U. Gundert-Remy, C. Lambré, J. Leblanc, O. Lindtner, P. Moldeus, P. Mosesso, A. Oskarsson, D. Parent-Massin, I. Stankovic, I. Waalkens-Berendsen, M. Wright, M. Younes, P. Tobback, S. Ioannidou, S. Tasiopoulou and R. A. Woutersen, *EFSA J.*, 2017, **15**, e04866.
25. NICNAS, *Public Report: Gelatins, reaction products with glutaraldehyde*, National Industrial Chemicals Notification and Assessment Scheme (NICNAS), 2019.
26. OECD, *SIDS Initial Assessment Report for Ammonium Perfluorooctanoate and Perfluorooctanoic Acid*, Organisation for Economic Cooperation and Development (OECD), Paris, France, 2008.
27. C. A. Ng and K. Hungerbühler, *Environ. Sci. Technol.*, 2014, **48**, 4637–4648.
28. Canadian Environmental Protection Act (CEPA), Categorization Results from the Canadian Domestic Substance List, https://canadachemicals.oecd.org/Search.aspx (accessed 1 November 2022).
29. OECD, *Screening Information Data Set for Ethylene, 74-85-1*, 1998.
30. E. T. Akhmetova, *Vestn. Oftalmol.*, 1977, **4**, 58–59.
31. International Agency for Research on Cancer (IARC), *IARC Monographs on the Evaluation of the Carcinogenic Risk of Chemicals to Humans Volume 19 (Polyethylene)*, 1987.
32. S. Atis, B. Tutluoglu, E. Levent, C. Ozturk, A. Tunaci, K. Sahin, A. Saral, I. Oktay, A. Kanik and B. Nemery, *Eur. Respir. J.*, 2005, **25**, 110–117.
33. International Agency for Research on Cancer (IARC), *IARC Summary & Evaluation, Volume 60 (Propylene)*, 1994.
34. Australian Industrial Chemicals Introduction Scheme (AICIS), *Chemicals unlikely to require further regulation to manage risks to environment – Evaluation statement*, 2022.
35. Australian Industrial Chemicals Introduction Scheme (AICIS), Assessment search: 9000-69-5 (Pectin), https://www.industrialchemicals.gov.au/chemical-information/search-assessments?assessmentcasnumber=9000-69-5%2C+ (accessed 21 November 2022).
36. K. H. Kirkland and K. D. Rosenman, *Occup. Environ. Med.*, 2022, **79**, 287.
37. Australian Industrial Chemicals Introduction Scheme (AICIS), Search assessments: 9000-70-8 (Gelatins), https://www.industrialchemicals.gov.au/chemical-information/search-assessments?assessmentcasnumber=9000-70-8+ (accessed 21 November 2022).
38. S. Farris, K. M. Schaich, L. Liu, P. H. Cooke, L. Piergiovanni and K. L. Yam, *Food Hydrocolloids*, 2011, **25**, 61–70.
39. J. Jin, M. Song and D. J. Hourston, *Biomacromolecules*, 2004, **5**, 162–168.
40. N. Cao, Y. Fu and J. He, *Food Hydrocolloids*, 2007, **21**, 575–584.
41. S. Woranuch, R. Yoksan and M. Akashi, *Carbohydr. Polym.*, 2015, **115**, 744–751.

42. B.-S. Chiou, R. J. Avena-Bustillos, P. J. Bechtel, H. Jafri, R. Narayan, S. H. Imam, G. M. Glenn and W. J. Orts, *Eur. Polym. J.*, 2008, **44**, 3748–3753.
43. A. Bigi, G. Cojazzi, S. Panzavolta, K. Rubini and N. Roveri, *Biomaterials*, 2001, **22**, 763–768.
44. J. Vartiainen and A. Harlin, *Mater. Sci. Appl.*, 2011, **02**, 346.
45. P. G. Gan, S. T. Sam, M. Omar and M. F. Abdullah, *IOP Conf. Ser.: Mater. Sci. Eng.*, 2020, **957**, 012038.
46. S. Farris, K. M. Schaich, L. Liu, L. Piergiovanni and K. L. Yam, *Trends Food Sci. Technol.*, 2009, **20**, 316–332.
47. A. Bigi, G. Cojazzi, S. Panzavolta, N. Roveri and K. Rubini, *Biomaterials*, 2002, **23**, 4827–4832.
48. D. P. Speer, M. Chvapil, C. D. Eskelson and J. Ulreich, *J. Biomed. Mater. Res.*, 1980, **14**, 753–764.
49. H. Katagiri, T. Yamamoto, A. Uchimura, M. Tsunoda, Y. Aizawa and H. Yamauchi, *Ind. Health*, 2011, **49**, 328–337.
50. L. L. Sano, A. M. Krueger and P. F. Landrum, *Aquat. Toxicol.*, 2005, **71**, 283–296.
51. Agency for Toxic Substances and Disease Registry (ATSDR), *Toxicological Profile for Glutaraldehyde*, 2017.
52. OECD, *Oecd Sids: Glutaraldehyde*, Unep Publications, 2001.
53. EU GHS Regulation (EC) No. 1272/2008, 2008.
54. TEDX, TEDX (The Endocrine Disruption eXchange) List of Potential Endocrine Disruptors, https://endocrinedisruption.org/interactive-tools/tedx-list-of-potential-endocrine-disruptors/search-the-tedx-list (accessed 20 October 2022).
55. K. Pal, A. T. Paulson and D. Rousseau, in *Modern Biopolymer Science*, ed. S. Kasapis, I. T. Norton and J. B. Ubbink, Academic Press, San Diego, 2009, pp. 519–557.
56. European Chemicals Agency (ECHA), 4-Hydroxy-3-methoxycinnamic acid, https://echa.europa.eu/information-on-chemicals/cl-inventory-database/-/discli/details/76092 (accessed 20 October 2022).
57. S. Tunç and O. Duman, *Appl. Clay Sci.*, 2010, **48**, 414–424.
58. A. Hiremath, A. A. Murthy, S. Thipperudrappa and K. N. Bharath, *Cogent Eng.*, 2021, **8**, 1991229.
59. J. George and Siddaramaiah, *Carbohydr. Polym.*, 2012, **87**, 2031–2037.
60. Y. Habibi, L. A. Lucia and O. J. Rojas, *Chem. Rev.*, 2010, **110**, 3479–3500.
61. T. Í. S. Oliveira, L. Zea-Redondo, G. K. Moates, N. Wellner, K. Cross, K. W. Waldron and H. M. C. Azeredo, *Food Chem.*, 2016, **198**, 107–112.
62. J. F. Martucci and R. A. Ruseckaite, *Polym.-Plast. Technol. Eng.*, 2010, **49**, 581–588.
63. R. Jeya Shakila, E. Jeevithan, A. Varatharajakumar, G. Jeyasekaran and D. Sukumar, *Food Chem.*, 2012, **135**, 2260–2267.
64. M. Abdollahi, M. Rezaei and G. Farzi, *J. Food Eng.*, 2012, **111**, 343–350.
65. M. Chaichi, M. Hashemi, F. Badii and A. Mohammadi, *Carbohydr. Polym.*, 2017, **157**, 167–175.

66. T. C. Mokhena, M. J. John, M. J. Mochane, A. Mtibe, T. S. Motsoeneng, T. H. Mokhothu and C. A. Tshifularo, in *Soil Microenvironment for Bioremediation and Polymer Production*, 2019, pp. 283–306.
67. H. Celebi and A. Kurt, *Carbohydr. Polym.*, 2015, **133**, 284–293.
68. A. Mtibe, T. C. Mokhena, T. H. Mokhothu and M. J. Mochane, in *Soil Microenvironment for Bioremediation and Polymer Production*, ed. N. Jamil, P. Kumar and R. Batool, John Wiley & Sons, 2019, pp. 319–338.
69. Australian Industrial Chemicals Introduction Scheme (AICIS), Assessment search: Montmorillonite, https://www.industrialchemicals.gov.au/chemical-information/search-assessments-keywords (accessed 27 October 2022).
70. N. Khatoon, M. Q. Chu and C. H. Zhou, *J. Mater. Chem. B*, 2020, **8**, 7335–7351.
71. Australian Industrial Chemicals Introduction Scheme (AICIS), Search assessments: 9004-34-6 (Cellulose), https://www.industrialchemicals.gov.au/chemical-information/search-assessments (accessed 27 October 2022).
72. A. R. Gibbs and F. D. Pooley, *Occup. Environ. Med.*, 1994, **51**, 644.
73. B. O'Connor, R. Berry and R. Goguen, in *Nanotechnology Environmental Health and Safety*, ed. M. S. Hull and D. M. Bowman, William Andrew Publishing, Oxford, 2nd edn, 2014, pp. 225–246.
74. A. Wicker, *Vox*, 2019.

CHAPTER 5

Alternatives to PFASs for the Surfactant Role in Floor Polish

JENNA A. TAN,*[a] TESSA A. WARDLE[b] AND YUNING XU[b]

[a] Department of Chemistry, University of California, Berkeley, CA, 94720, USA; [b] University of California, Berkeley, School of Public Health, Berkeley, CA, 94704, USA
*Email: jatan@berkeley.edu

5.1 Challenge Overview

This chapter examines potential alternatives to perfluoroalkyl and polyfluoroalkyl substances (PFASs) used in floor polish in response to the 2021 passage of the Maine Legislature bill, "An Act To Stop Perfluoroalkyl and Polyfluoroalkyl Substances Pollution" (LD 1503), which includes banning PFASs in non-essential use product categories, such as floor polish, by 2030. This chapter discusses the potential of rhamnolipids and sodium lauroyl glutamate as alternatives to PFASs in floor polish products with rhamnolipids as the leading contender. The potential of rhamnolipids and sodium lauroyl glutamate in floor polish can be further clarified in future studies that compare the final performance of the two biosurfactants on relevant floor surfaces and improve understanding of aquatic toxicity and long-term effects on human health. Although the chapter focuses on alternatives to PFAS use in floor polishes, it is valuable to explore strategies beyond PFAS replacement to improve general hazard concerns surrounding floor polish formulations.

The prevalence of PFASs in drinking water systems has emerged as a major concern, since the complete removal and destruction of PFASs is very difficult and often expensive. In Maine, testing conducted between

Green Chemistry Series No. 81
Toward a PFAS-free Future: Safer Alternatives to Forever Chemicals
Edited by Simona A. Bălan, Thomas A. Bruton and Kimberly G. Hazard

Published by the Royal Society of Chemistry, www.rsc.org

2016–2020 revealed high levels of PFASs in drinking water, particularly residential well water and some school water sources.[1–3] In part influenced by this concerning discovery, the Maine Legislature passed laws in 2021 requiring the removal of PFASs from non-essential use in products sold in the state by 2030.[4] This regulation spurred Defend Our Health, a Maine-based non-profit that works to protect residents from the harmful effects of toxic chemicals, to compile recommendations for products in which PFASs are considered "avoidable" components.

This project was conducted in partnership with Defend Our Health to explore the existence of viable alternatives to using PFASs in floor polish, therefore making the use of PFASs in floor polish formulations "avoidable," as part of the UC Berkeley graduate course, Greener Solutions (August to December 2021). Course instructors (Dr Megan Schwarzman, Dr William Hart-Cooper, and Ms Kimberly Hazard) facilitated and guided the partnership between Patrick MacRoy, Deputy Director at Defend Our Health, and the student team, composed of Yuning Xu (environmental health sciences), Tessa Wardle (environmental health sciences), and Jenna Tan (chemistry).

PFASs in floor polish pose a relevant concern for drinking water contamination, especially well water sources.[5] According to personal correspondence with a floor polish expert from SC Johnson, the primary method of floor polish application, when considering all possible settings, is with a mop and bucket.[6] Washing the floor polish application tools (*i.e.*, mop and bucket) could lead to residual floor polish going down the drain and ending up in wastewater, where removing PFASs is challenging. PFASs can remain in the treated water and solid waste (*i.e.*, sludge) upon release into the environment and contaminate surrounding areas including groundwater sources.[5,7]

Although assessing accurate PFAS concentrations across floor polish formulations is difficult, PFASs are pervasive in the floor polish industry. According to a 2008 survey conducted by the Danish EPA, DuPont claimed that almost every floor polish formulation on the market contained a fluorosurfactant (100–500 ppm, 0.01–0.05%).[8] This was further corroborated in a recent report conducted by the UK Environment Agency, in which PFASs were reported to be used in four floor polish products as a wetting and leveling agent with concentrations ranging from 0.015 to 1.448%.[9]

Current large-scale users of floor polish include, but are not limited to, schools, hospitals, and large retailers including grocery stores. Floor polish primarily protects and extends the lifespan of floors, and secondarily enhances the visual appearance of floors and facilitates easier cleaning.[10] Floor polish is also marketed to residential consumers for use on home floors.

PFASs act as surfactants in floor polish formulations. Surfactants are characterized by their amphiphilic nature, having an affinity for both water (hydrophilic) and fats (lipophilic or hydrophobic). All fluorosurfactants share a hydrophobic fluorine–carbon tail which is responsible for their surfactant behavior. Fluorosurfactants aid in spreading the floor polish across the floor, a characteristic called wetting, and help ensure that no visual impurities are present when the polish dries, a characteristic called leveling.[11] Typically, floor polish application requires four to five coatings to ensure full performance.[6] Fluorosurfactants excel as floor polish additives for the desired property of recoatability. Although PFAS replacements perform well for the first one or two coats, their performance tends to decline when more than four coats are applied.[6] Fluorosurfactants facilitate the application of four to five coats without compromising gloss and high leveling performance while resisting dark scuff marks and dents.[11] Additionally, fluorosurfactants are effective in small doses (usually 100–275 ppm),[6] which helps ensure that other properties of the floor polish do not start to degrade. A small fluorosurfactant concentration also means that they are extremely versatile for use in many different formulations and under variable conditions.[6]

Potassium *N*-ethyl perfluoro alkane sulfonamidoacetate (*N*-EtFASAA) is a group of PFASs that represents the kinds of fluorosurfactants commonly used in floor polish. One type of *N*-EtFASAA that we focus on in this chapter is 2-(*N*-ethylperfluorooctanesulfonamido) acetic acid, referred to as *N*-EtFOSAA (CAS no. 2991-50-6). *N*-EtFOSAA is highly persistent, can bioaccumulate, and is associated with toxic health effects to humans.[12–15] Although there are still a number of data gaps regarding how *N*-EtFOSAA affects other important health endpoints, its highly-studied degradation product, perfluorooctane sulfonic acid (PFOS), offers potential insight into the data gaps. PFOS is highly stable in the environment and does not degrade.[16] Various harmful health effects are associated with PFOS, including endocrine disruption, carcinogenicity, organ toxicity, immunotoxicity, neurotoxicity, and more.[17] Perfluorobutanesulfonic acid (PFBS) is another important degradation product to consider from the *N*-EtFASAA family and shares many of these concerns, additionally presenting high concern as an eye and skin irritant and a reproductive toxicant.[18]

Identifying alternatives to PFASs in floor polish is an ongoing research challenge. PFAS alternatives must act as wetting agents and facilitate the spread of floor polish across a floor surface, as well as concurrently level the polish. Alternative non-fluorinated floor polish formulations are available on the market, some of which use essential oils or naturally derived wax. However, they still have not captured a large portion of the floor polish market since they may not perform as well as PFAS-based formulations or require more labor during application. Other alternatives such as non-fluorinated gemini surfactants, which are two surfactant molecules that are linked together, are emerging for use in floor polish formulations and can have similar or better performance than traditional fluorosurfactants.[19,20] Alkyl polyglucosides and alkyl ethoxy polyglucosides are also promising non-ionic biosurfactants that demonstrate biodegradation and strong detergency for use in cleaning applications and may be considered for floor polish formulations.[21–23] Meanwhile, patents for formulations that are not yet on the market include non-fluorinated water-based floor wax, but it is unclear as to when this material will become widely available.[24] Based on the currently available research information, rhamnolipids and amino acid surfactants, specifically sodium lauroyl glutamate, both demonstrate strong technical potential as leveling agents while being more biodegradable with lower health risks.

Small businesses and individual consumers may not always be aware of the presence of PFASs in floor polish or the associated health risks, as this issue has not received the same level of attention as the presence of PFASs in other consumer products, such as non-stick pans or water-repellent outerwear. Thus, it is important that state governments and non-profit organizations, such as Defend Our Health, implement initiatives to protect consumer health. Initiatives include mandatory labelling of PFASs in products, providing frameworks for reducing PFAS pollution, or removing all products with non-essential use of PFASs from state contracts as an effort "to model environmentally-friendly supply chain practices."[25,26]

5.1.1 Problem Statement

The aim of this project was to investigate the use and disposal of floor polish products that contained PFASs and find safer functional chemical alternatives. Identifying such an alternative would allow the manufacturing of floor polishes without the addition of PFASs, and the Maine legislature could ban the use of PFASs in floor polish products sold in Maine. Ultimately, removing PFASs from floor polish would contribute to the protection of the health of humans and the environment.

5.1.2 Approach

The goal of this project was to find viable candidates with a strong baseline performance and improved human and environmental health outcomes relative to fluorosurfactants used in floor polish. The scope of this work focused on alternative surfactants that were commercially available. The technical performance of the proposed solutions was assessed alongside the potential health and environmental impacts.

5.1.2.1 Technical Approach

PFASs serve multiple functions in floor polish. As such, the search criteria for replacements were narrowed down to the essential functions, wetting and leveling performance. Both wetting and leveling agents help the floor polish to spread evenly over a substrate in different ways. The molecular interactions and interfacial energies involved in the spreading of a liquid were simplified in order to generalize and evaluate the wetting and leveling potential of alternative candidates as shown in Figure 5.1. Wettability is related to the liquid–substrate interface and whether a liquid droplet will spread (Figure 5.1).[27] This phenomenon can be measured using the contact angle (θ), where a contact angle that is 90° or higher indicates low wetting.[28]

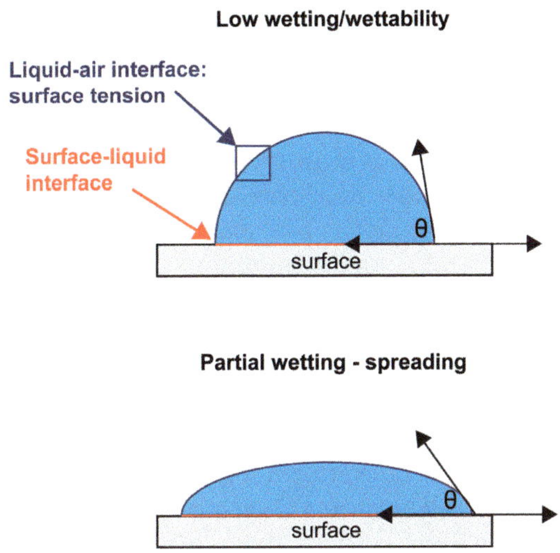

Figure 5.1 Schematic of a droplet wetting on substrate illustrating low and partial wetting[27] that also distinguishes surface tension from wettability. Reproduced from ref. 27 with permission from the Royal Society of Chemistry.

When the contact angle is less than 90°, it indicates that there is partial or high wetting, which leads to spreading.[28] In this work, the criterion for good wettability was set as a contact angle measurement of less than 90° and best wettability was assigned to contact angle values of less than 45°, which was a halfway point toward what is considered complete wetting at 5°.[28] This distinction served as a rough metric to provide more clarity on wettability performance for this case study.

Leveling occurs at the liquid–air interface and can be measured through surface tension (Figure 5.1). Surface tension refers to the energy required to increase the surface area of a liquid, or more generally, the force required to keep a liquid together as a droplet.[29–31] A surfactant's ability to reduce surface tension is measured by how well it reduces the surface tension of water, which has one of the highest surface tensions for a liquid at 72 mN m^{-1}.[31] Excellent leveling agents need to reduce the surface tension of water by more than 50% (less than 36 mN m^{-1}), which is the value used as a leveling performance criterion in this work.

Additional technical metrics that aided in the assessment of surfactant candidates for this work were the critical micelle concentration (CMC) and the octanol–water partition coefficient (K_{ow}). CMC is the concentration at which surface tension stops changing with surfactant concentration. It is a helpful benchmark for comparing the amount of surfactant that is needed to achieve minimum surface tension, where a small CMC value indicates that less concentration of surfactant is needed.[31,32] Since most floor polishes contain between 100–275 ppm, or 2–7 mM, of PFASs, a competitive surfactant should have a comparable CMC.[6] The K_{ow} is used to measure whether a substance has greater affinity to octanol, an oil or fatty phase, or water, which is a good indicator of a substance's ability to repel oil, known as oleophobicity.[33] The strong oleophobicity of fluorosurfactants potentially drives a floor polish formulation's tolerance for extensive recoating.[10] Thus, the K_{ow} serves as a proxy for recoating potential in this chapter. K_{ow} is a unitless value commonly reported in the logarithmic form ($\log K_{ow}$) since K_{ow} spans many orders of magnitude. $\log K_{ow}$ spans from −3 (very hydrophilic) to +10 (very hydrophobic).[33] An ideal candidate would be an oleophobic, or hydrophilic, compound, which means that smaller values indicate better recoating potential. The upper $\log K_{ow}$ threshold at which moderate to optimal efficiency could be achieved was set as 5, since this is a common designation used to determine whether a substance is oleophilic *versus* oleophobic.[34] Values above 5 indicate an oleophilic compound that may demonstrate poor recoating ability.

PFASs often technically outperform other types of surfactants due to the influence of the strong carbon–fluorine chain;[35,36] therefore, sodium dodecyl sulfate (SDS) was used to provide a technical baseline performance for non-fluorinated surfactants in this work. SDS is a commercially available anionic surfactant commonly used in cleaning and hygiene products. Figure 5.2 summarizes the technical metrics used in this chapter devised by the student group.

5.1.2.2 GreenScreen Score Assignment for Health and Environmental Endpoints

The Pharos Project is an online database that provides overviews of existing chemical hazard data. Pharos provided initial information from authoritative

Performance Metrics	High Efficiency (H)	Moderate Efficiency (M)	Low Efficiency (L)
Wettability	Water contact angle < 45°	45° ≤ Water contact angle ≤ 90°	Water contact angle > 90°
Leveling	Surface tension < 36 mN m^{-1}	36 mN/m ≤ Surface tension ≤ 54 mN m^{-1}	Surface tension > 54 mN m^{-1}
Surfactant Concentration	CMC < 2 – 7 mM	7 mM ≤ CMC ≤ 14 mM	CMC > 14 mM
Recoatability	$\mathrm{Log}K_{ow}$ < 2	2 ≤ $\mathrm{Log}K_{ow}$ ≤ 5	$\mathrm{Log}K_{ow}$ > 5

Figure 5.2 Summary of technical metrics for predicting surfactant performance in floor polish applications.

lists to evaluate the health and environmental endpoints for existing PFASs used in floor polish, their degradation products (*N*-EtFOSAA, PFBS, and PFOS), as well as proposed alternatives. Gaps in the information were filled by reviewing the literature and using CompTox calculations, which helped model chemical hazard concerns based on predictions or the hazards of similarly structured chemicals. The initial evaluation was standardized using the Globally Harmonized System (GHS) of Classification and Labeling of Chemicals. The GHS scores were converted into GreenScreen assignments using a procedure adapted from GreenScreen for Safer Chemicals.[37] This chapter focuses on the human and environmental endpoints that were most relevant in the entire life cycle of the PFAS alternatives. For example, when considering the disposal of the alternative candidates, criteria for success in this part of the life cycle included biodegradability and low environmental persistence, particularly in water. The hazard potential of the endpoints was denoted by a score of high or very high (red), medium (yellow), or low (green), where the confidence in the score was described as low (italicized) or high (bolded) depending on whether studies directly or indirectly described the substance or endpoint of interest. A data gap was assigned to an endpoint if there was a complete lack of studies directly related to the substance and the endpoint of interest.

5.1.3 Partners Involved in This Work

The Maine-based non-profit Defend Our Health advocates for specific bans on non-essential uses of PFASs in product categories where such a ban could have a big impact in terms of protecting health, and plans to provide the Maine legislature with suggestions of such categories.[2] Floor polish is one category of interest because it could be argued that PFAS use in floor polish is not essential if sufficient alternatives to PFASs exist and if the benefits of PFAS use in this product do not outweigh the risks. By identifying appropriate alternatives to PFASs in floor polish or providing existing safer floor polish options, Defend Our Health's goal to ban non-essential uses of PFASs would be easier to accomplish and could be enacted quickly under the 2021 Maine Act To Stop Perfluoroalkyl and Polyfluoroalkyl Substances Pollution.[2,3] The student group was in email and video correspondence with the deputy director in 2021.

The student group also had personal communication with a floor polish expert at SC Johnson who provided information about their past research on PFAS alternatives in floor polish and current performance standards for floor polish through interviews and a presentation. This information was invaluable for guiding the approach to the project as well as providing key background on the existing use of PFASs in floor polish.

5.2 Solutions

Surfactants in floor polish solutions need to work at two interfaces: between the floor polish solution and air, and between the floor polish solution and

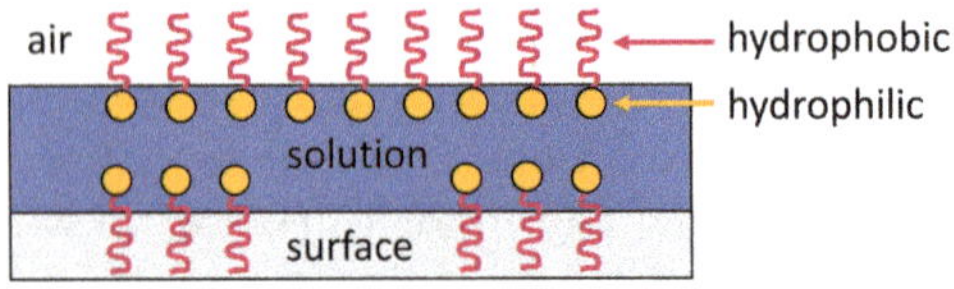

Figure 5.3 Schematic of how surfactant molecules orient themselves in solution. For a water-based solution, the hydrophilic end of the surfactant orients toward the solution, and the hydrophobic end orients away from the solution.

the floor surface (Figure 5.3). In both cases, the hydrophilic component orients toward the floor polish solution (assuming it is water-based), while the lipophilic (or hydrophobic) component orients away from the bulk floor polish solution and toward either the air or the floor surface. The technical performance of two bio-based non-fluorinated surfactants was evaluated against the two most popular fluorosurfactants currently used in floor polish formulations that go by the trade names FS-60 and FS-65, as well as a baseline non-fluorinated surfactant, SDS. For data gaps with FS-60 and FS-65, other fluorosurfactant data were used.

5.2.1 Rhamnolipids

Microorganisms produce biosurfactants in response to physical and chemical stressors to advantageously interact with and change the surface of their local environments.[38,39] Biosurfactants have garnered widespread interest for their applications in emulsification, cleansing, foaming, and dispersion in a variety of industries including agriculture, pharmaceuticals, food, and cosmetics.[40–43] Appealingly, biosurfactants can degrade under aerobic and anaerobic conditions, meaning that they are unlikely to persist in the environment.[44,45]

As one of the leading commercial biosurfactants, rhamnolipids have the potential to act as a leveling and wetting agent in floor polish.[40] The bacterial species *Pseudomonas (P.) aeruginosa* produces rhamnolipids, which are composed of one or two sugar molecules and a fatty acid chain (Figure 5.4).[45] Different feedstocks can serve as the substrate for microbial fermentation (*e.g.*, waste cooking oil, fats, oils, and sugars), which could lead to the creation a closed production loop and different conditions whereby rhamnolipid physicochemical properties could be manipulated.[44,45]

5.2.1.1 *Technical Performance of Rhamnolipids*

Mono-rhamnolipids demonstrated strong leveling potential with surface tension values comparable to the fluorosurfactants (18–19 mN m^{-1}) currently used in floor polish (Figure 5.6).[46,47] They also displayed a CMC only two times larger than those of the fluorosurfactants (0.23 mM for proxy compound Capstone 1157)[48] and around 20 times smaller than the CMC of SDS. Rhamnolipids

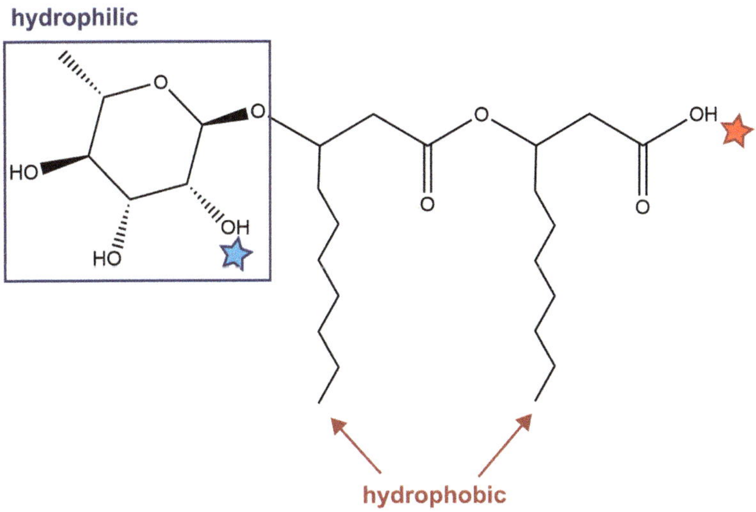

Figure 5.4 A mono-rhamnolipid molecule, with the rhamnose molecule (sugar) enclosed in the blue box and the fatty acid chains (hydrophobic ends) indicated by arrows.

reduced the surface tension to a range of 26–29 mN m^{-1},[45] which was comparable to that of SDS (24–35 mN m^{-1}),[49] and noteworthy given that the CMC value for rhamnolipids (0.4 mM)[50] was much smaller than the CMC for SDS (8 mM),[42] indicating that a relatively low surface tension was achieved with a smaller amount of rhamnolipids. SDS exhibited multiple CMC values, but the lowest was 8 mM.[42]

Since the contact angle strongly depends on substrate material, a fair evaluation of wettability potential required comparing water contact angle measurements across the same substrates, such as polyvinyl chloride (PVC). Based on the information available, only the wettability potential between SDS and rhamnolipids on PVC was directly comparable. In a study evaluating the wettability of rhamnolipids, it was observed that they displayed smaller contact angles at almost all concentrations compared to SDS, suggesting that rhamnolipids are better wetting agents than SDS, but the contact angles vary widely from 20–45°.[51] Although rhamnolipids achieved a low contact angle of 20° indicating excellent wettability, this required a concentration of approximately 1.5 mM, which was about three times larger than the rhamnolipid's CMC of 0.4 mM[50] and may contribute to formulation costs. Rhamnolipids may perform well as both a wetting and leveling agent as indicated by their high-performing surface tension and contact angle values on PVC; however, testing will be required to find the rhamnolipid concentration that optimizes both functions.

The CompTox Chemicals Dashboard from the U.S. Environmental Protection Agency (EPA) provided estimates for the $\log K_{ow}$ values of rhamnolipids (CAS no. 4348-76-9) and a proxy PFAS for FS-60 and FS-65 (*N*-EtFOSAA,

CAS no. 2991-50-6).[52] Rhamnolipids were predicted to have a $\log K_{ow}$ of 5.05[53] (meaning more hydrophobic or oleophilic), suggesting that rhamnolipids favor fatty environments. Since this work used oleophobicity to predict a rhamnolipid's ability to assist in recoating, its higher $\log K_{ow}$ indicated possibly poor recoating ability. However, this value is just an imperfect proxy for measuring recoating ability and should be interpreted with caution.

5.2.1.2 *Strategy for Tuning Rhamnolipid Performance*

There are strategies to fine tune the surfactant properties for improved performance, such as changing the salt counterion or concentration. For example, past experiments on PFOA as a surfactant showed that PFOA without a salt yielded an effective surface tension that was less than half of the value of PFOA with a sodium counterion.[54] The same study showed that PFOS with a potassium salt exhibited the lowest surface tension compared to other salt counterions (*e.g.*, lithium and ammonium).[54] Meanwhile, varying the concentration of lithium chloride mixed with SDS manipulated the resulting contact angle of SDS.[55] There have been studies exploring the impact of salt concentration on a rhamnolipid's ability to lower surface tension.[50,56] For mixtures of rhamnolipids with different salt (NaCl) concentrations, surface tension generally decreased as the salt concentration increased.[50] In particular, increasing the NaCl concentration up to 40 $g\,L^{-1}$ consistently decreased the overall range of surface tension values.[50] This systematic change in NaCl concentration offers a method for manipulating the rhamnolipid surfactant properties.

5.2.1.3 *Health and Environmental Performance of Rhamnolipids in Floor Polish*

A limited life cycle analysis of rhamnolipids was conducted in the context of a floor polish product, in order to evaluate their environmental and health outcomes compared to those of PFASs (Figure 5.7).

5.2.1.4 *Rhamnolipids in Floor Polish Production*

Rhamnolipids demonstrated a lower risk compared to PFASs in floor polish products across six hazard categories, as denoted by a low hazard score in the hazard table (Figure 5.7). In particular, there was convincing evidence that most endpoints related to human health outcomes have a low level of concern despite the current data gaps. Patowry *et al.* tested the cytotoxicity of rhamnolipids in mouse cell lines and did not find any cytotoxic effects.[57] In fact, rhamnolipids have been tested as a natural anticancer agent, with one research study finding promise in rhamnolipids' ability to assist a traditional chemotherapy drug to induce damage or death in cancer cells.[58] Overall, rhamnolipids presented little concern for production workers who may be exposed to rhamnolipids through dermal contact or inhalation.

While rhamnolipids presented a low concern in floor polish production, their source material, *P. aeruginosa,* presented possible risks during rhamnolipid production. *P. aeruginosa* is an opportunistic infectious bacteria that produces a large number of extracellular toxins.[59] Although there is a potential health risk of occupational exposure to *P. aeruginosa* during rhamnolipid production, this risk can be reduced through appropriate hygiene protocols and personal protective equipment use.[60] Furthermore, research on expressing enzymes for rhamnolipid production in non-pathogenic hosts will also decrease occupational exposure risk.[61]

Another concern associated with the production of rhamnolipids is related to the use of hazardous solvents such as chloroform. Such solvents are used to increase the yield in rhamnolipid production, and may lead to potential employee exposure risks.[62] Replacing problematic solvents with greener solvents and optimizing the rhamnolipid extraction and purification process will reduce both the reliance on and risk from hazardous solvents.[63,64] Despite the presence of human health concerns during the production of rhamnolipids, there exist many strategies, such as those described above, to address these concerns.

5.2.1.5 Rhamnolipids in Floor Polish Application

During floor polish application, there is a risk of both acute and chronic exposure. Acute exposure can occur from any accidental spills, while chronic exposure could result from repeated contact with floor polish through either dermal exposure or inhalation. Floor waxing workers are particularly at risk, and can be exposed to various PFASs from volatilized particles or dermal exposure during floor polish application.[7] This risk likely can be extrapolated to include those who apply floor polish at home, as well as those who are in the same airspace when the polish is being applied. While safety policies can mitigate PFAS exposure for both floor waxing staff and surrounding people (*e.g.*, personal protective equipment, school application when no children are present), using alternative surfactants such as rhamnolipids drastically reduces these hazards.[43,44] Rhamnolipids displayed low hazard concern in most of the endpoints considered, including carcinogenicity, mammalian toxicity, systemic toxicity, respiratory toxicity, and skin irritation. However, rhamnolipids presented a higher risk of eye irritation,[65,66] which can be mitigated through the use of personal protective equipment.

5.2.1.6 Rhamnolipids in Floor Polish Disposal

As mentioned previously in the chapter, spent floor polish may enter water systems, which raises concerns regarding aquatic toxicity and persistence. Rhamnolipids biodegraded under both aerobic and anaerobic conditions, so they likely would be removed by wastewater treatment plants[67] and present a lower risk for unforeseen downstream consequences. Furthermore, rhamnolipids enhanced the biodegradation of other pollutants, removed heavy metals, and also helped wet soil to improve irrigation, in contrast to harmful PFAS contamination of the soil.[45,68–70] One remaining area of concern was the potential of rhamnolipids to inhibit algae growth and disrupt soil microbial activity.[71] Finally, there was a minor concern regarding the bioaccumulation potential of rhamnolipids;[53] however, this is based on a modeled estimate using the log K_{ow} value, so experimental data are needed in order to draw stronger conclusions.[33,72]

The Danish EPA has categorized rhamnolipids as category 2 (using GHS criteria) for chronic aquatic toxicity. This classification, which is based on modeled results, means that rhamnolipids are toxic to aquatic life and exhibit long-lasting effects.[73] Johan *et al.*, however, conducted a full battery of ecotoxicity tests of mono-rhamnolipids and found low acute toxicity based on fish embryo acute toxicity testing.[74] Additionally, they observed no mutagenic or estrogenic effects based on the Ames fluctuation assay and *in silico* modeling. In this study, rhamnolipids exhibited lower aquatic toxicity than synthetic surfactants, but slightly higher toxicity than other biosurfactants based on immobility testing in daphnids.[74] Based on the expected concentration of rhamnolipids that could end up in waterways, it was concluded that there is minimal toxicity concern.

Overall, while there is currently a lack of literature concerning the potential health and environmental hazards of rhamnolipids, there is enough evidence to suggest that this alternative is safer than the baseline surfactants.

5.2.2 Amino Acid Surfactants – Sodium Lauroyl Glutamate

Amino acid surfactants are biodegradable commercial surfactants currently used in industries such as cosmetics, cleaning products, and biomedical applications.[75] Amino acids serve as the building blocks of proteins in all living organisms, and their basic structures contain a carboxyl group, an amine group, and a distinct functional group that endows each amino acid with unique chemical properties, such as charge (*e.g.*, anionic and cationic).[75,76] Amino acid surfactants form through the condensation of natural amino acids with fatty acids derived from an oleochemical source.[1] This chapter focuses on anionic surfactants, specifically sodium lauroyl glutamate (SLG) whose amino acid precursor is glutamic acid, since its negative charge allows the molecule to effectively lift and suspend dirt, which is a desirable property for floor polishes.[77]

SLG demonstrates high surface activity, low toxicity, and is predicted to have quick biodegradation based on its relatively simple structure (Figure 5.5).[76] Although current production methods for amino acid surfactants rely on solvents to increase yield at a lower cost, there is potential for greener production methods through the use of enzyme catalysis.[78] Additionally, exploration into more sustainable surfactant production focuses on using renewable and raw feedstocks like waste proteins and waste cooking oils to form the fatty acid chain precursor.[45]

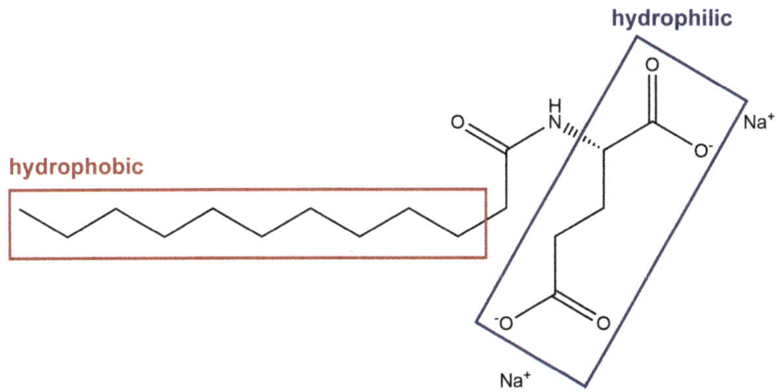

Figure 5.5 The sodium lauroyl glutamate molecule, with the hydrophilic component enclosed in the blue box, and the fatty acid chain (the hydrophobic end) outlined in the red box.

Technical Performance	Water contact angle (°) for wettability	Surface tension (mN m^{-1}) for levelling	Critical micelle concentration (mM) for surfactant concentration	Octanol-water partition coefficient (LogK$_{OW}$) for oleophobicity
Fluorosurfactants	~51-56° for proxy substances FS-61 and FS-63 (*M*)[98]	19 at 0.05% for FS-60 and FS-65 (**H**)[46,47]	0.23 for Capstone 1157 (*H*)[48]	7.35 (5.31-9.34) for N-EtFOSAA (*L*)[52]
Sodium dodecyl sulfate (SDS)	~20° at 3.48 mM on PVC (*H*)[51]	23.8-34.6 (*H*)[49]	8-8.5 (M)[42]	1.69 (*H*)[72]
Rhamnolipids	~20° at 1.5 mM on PVC (*M*)[51]	26-29 (*H*)[45]	0.4 (*H*)[50]	5.05 (1.50-7.99) (*L*)[53,65]
Sodium lauroyl glutamate	DG	38 at pH = 7 (*M*)[79]	3.1 (*H*)[79]	< 3 (*H*)[80]

Key

Efficiency	High (H)	Moderate (M)	Low (L)	Data Gap (DG)
Confidence	*Low*	**High**		

Figure 5.6 Comparative technical summary between fluorosurfactants (FS-60 and FS-65),[46–48,52,98] baseline non-fluorinated surfactant (SDS),[42,49,51,72] rhamnolipids,[45,50,51,65] and sodium lauroyl glutamate.[79,80]

5.2.2.1 *Sodium Lauroyl Glutamate Technical Performance*

SLG reduced surface tension to about 38 mN m^{-1}, effectively reducing the surface tension of water (72 mN m^{-1}) by approximately 50% (Figure 5.6).[79] Based on conversations with a floor polish expert, it was noted that although fluorosurfactants achieve the lowest surface tension,[46,47] this property alone does not consistently predict the overall leveling performance.[6] Thus, SLG likely has strong potential as a leveling agent in floor polishes given that it meets this project's metric for moderate surface tension performance and is comparable to SDS (Figure 5.6). Furthermore, the concentration of SLG needed to achieve the minimum surface tension, measured using CMC, was very low (3.1 mM).[79] Comparing the wettability of SLG to the baseline surfactants was difficult due to the lack of water contact angle studies. SLG has a reported $\log K_{ow}$ of less than 3, which means that it is very oleophobic and suggests that it has good recoatability potential.[80]

5.2.2.2 *Strategy for Tuning Amino Acid Surfactant Performance*

One approach to tune the properties of SLG and other amino acid surfactants is to link two identical surfactant molecules together with a spacer chain to synthesize a compound called an amino acid gemini surfactant.[75,81] They have not been widely used in commercial applications yet, and a gemini surfactant with SLG base units has not yet been synthetically explored. There is, however, growing research on amino acid gemini surfactants since they offer unique benefits to traditional gemini surfactants, including the ability to be manufactured from renewable resources and possibly low ecotoxicity and cytotoxicity.[81]

Gemini surfactants exhibit critical micelle concentrations up to two orders of magnitude lower than their single surfactant counterparts.[82] This means that a smaller concentration of a gemini surfactant is needed to achieve its lowest surface tension, which can result in production cost savings. However, their impact on absolute surface tension values is variable and dependent on the base surfactant molecule.[82,83] While gemini surfactants more efficiently reduce surface tension, their overall surface tension values at CMCs still range from 30–34 mN m^{-1}, which is not significantly different from our proposed SLG surfactant.[81,84,85] Although gemini surfactants cannot achieve a surface tension as low as a fluorosurfactant, surface tension alone is not predictive of the final performance of amino acid gemini surfactants in floor polish.

Other modifications to the gemini surfactant can further tune its surfactant properties. Some gemini surfactants with a short chain exhibit high viscosity at a low concentration, which may be undesirable for floor polish applications.[82] A study on the length of the hydrophobic fatty acid chain illustrated that increasing the carbon length from 6 to 14 carbons corresponds with a decrease in the critical micelle concentration to achieve their lowest surface tension values.[84]

While there are concerns that the increased chemical stability of gemini surfactants may result in poorer biodegradation, a growing number of studies focus on incorporating biodegradability into these molecules. Gemini surfactants that are sourced from arginine show better biodegradation and lower aquatic toxicity than bis-quaternary ammonium halide surfactants and monomeric counterparts.[75,81] Additionally, synthetically incorporating a spacer chain that is easily breakable reduces persistence.[86]

Overall, gemini surfactants have garnered significant attention due to the drastic changes in their surfactant properties achieved by combining two single surfactant units together. Most gemini surfactants that have been investigated are cationic due to easier synthetic manipulation; however, exploration of anionic gemini surfactants holds potential for growth.[84,86] In general, most gemini surfactants have been studied for their micelle formation[83,84] for application in cleaning applications (*e.g.*, residential, pharmaceuticals, oil recovery),[81,83] and there is opportunity for further studies on other surfactant properties that may be more relevant for floor polish applications.

5.2.2.3 *Health and Environmental Performance of Sodium Lauroyl Glutamate in Floor Polish*

Similar to the assessment of rhamnolipids, the health and environmental impacts of using SLG in floor polish were considered in a limited life cycle assessment of the product, from production to use to disposal (Figure 5.7). Sodium lauroyl glutamate appeared to be safer than PFASs at each stage of the life cycle, which is consistent with its listing on the EPA's Safer Chemical Ingredients List.[87]

5.2.2.4 *Sodium Lauroyl Glutamate in Floor Polish Production*

The European Chemicals Agency (ECHA) identified possibly high eye irritation potential with SLG, which could be a concern for production workers who may be exposed to higher levels of SLG.[80,88] There was also concern of

Hazard Assessment	C	M	R/D	E	AT	ST	N	SnR	IrS	IrE	AA/CA	P	B
N-EtFOSAA	*M*[12,100,101]	DG[102,103]	DG	DG[104–108]	*M*[12]	DG	DG[109–115]	DG	DG	DG[116]	DG	DG[13]	*H*[14,15]
PFBS	*M*[117,118]	*L*[102,119–122]	*H*[123–129]	*H*[18]	*M*[18]	**H**[130]	DG[131–135]	DG[135–137]	*VH*[18]	**H**[18]	*M*[18]	*H*[18]	*L*[14]
PFOS	**H**[16]	*L*[102,130,138]	**H**[16,102]	*H*[16]	**M**[16]	**H**[16]	*M*[134,139–141]	DG[142]	*M*[143]	*M*[16]	**H**[16]	**VH**[16]	**VH**[16]
Rhamnolipids	*L*[57,58]	*L*[74]	DG	DG[74]	*L*[53]	DG	DG[144]	DG	*L*[66]	*VH*[65,66]	**L**[73,74]	**L**[65,66]	*L*[53]
Sodium lauroyl glutamate (SLG)	DG	*L*[90,145]	*M*[89]	DG[146]	*L*[99]	DG	DG	*L*[90]	*H*[89]	*H*[89]	*L*[99]	DG	DG

Key	Hazard	High / Very High (H) or (VH)	Moderate (M)	Low (L)	Data Gap (DG)
	Confidence	*Low*	**High**		

Endpoint Abbreviations: Carcinogenicity (C), mutagenicity (M), reproductive and developmental toxicity (R/D), endocrine activity (E), acute mammalian toxicity (AT), systemic toxicity (ST), neurotoxicity (N), respiratory sensitization (SnR), skin irritation (IrS), eye irritation (IrE), acute and chronic aquatic toxicity (AA/CA), persistence (P), bioaccumulation (B)

Figure 5.7 Hazard assessment of three fluorosurfactants (*N*-EtFOSAA,[12–15,100–116] PFBS,[14,18,102,117–137] PFOS[16,102,130,134,138–143]) and two alternative surfactant candidates, rhamnolipids[53,57,58,65,66,73,74,144] and sodium lauroyl glutamate (SLG).[89,90,99,145,146]

skin irritation according to a submission by a manufacturer for the EU REACH hazard list,[89] with the degree of skin irritation shown to be dose dependent.[90] Surfactants tend to cause some degree of irritation because they reduce the surface tension and disrupt the cell's lipid membrane, so we suggest protocols for personal protective equipment to minimize skin and eye irritation.[91] Further identification of relevant concentrations will also determine safety protocols, as there are some conflicting studies and statements supporting SLG's low irritation potential.[89,90]

Group 1 endpoints outlined in Figure 5.7 (*i.e.*, carcinogenicity, mutagenicity, reproductive and developmental toxicity, and endocrine activity) are important to consider for the health of production workers since they often experience chronic exposure at varying levels to the substances they produce. There is potential concern over the reproductive and developmental toxicity of SLG according to the Danish EPA Advisory List.[89] This evaluation, however, was based on a modeled result,[89] so more experimental studies on SLG's impact on reproductive and developmental toxicity are needed. Fortunately, SLG exhibited low concern as a mutagen since it did not show reactivity to the Ames test, which may translate into low carcinogenic potential for production workers.[90]

5.2.2.5 Sodium Lauroyl Glutamate in Floor Polish Application

Routes of exposure during floor polish application include dermal contact and inhalation. Skin and eye irritation are a potential concern for the person applying the floor polish; however, these hazard endpoints are less concerning during the application than production, as production workers are more likely to encounter higher concentrations that could induce severe skin and eye irritation. Since SLG is used safely in cosmetics,[90] small concentrations likely do not lead to severe irritation.

There is concern over both acute effects from any accidental exposure during the use phase, as well as chronic, low-level exposure. SLG initially appears to not have significant concerns for group I endpoints that can be associated with chronic exposure, as seen in the hazard table (Figure 5.7). However, there are also substantial data gaps regarding the long-term human health impacts from exposure to SLG. Further research that clearly defines and evaluates the safety or hazards of SLG will be valuable.

5.2.2.6 Sodium Lauroyl Glutamate in Floor Polish Disposal

Aquatic toxicity, persistence, and bioaccumulation are important to consider given the possibility of floor polish materials getting washed down the drain. SLG provides a substantial improvement over PFASs because SLG readily biodegrades and poses a low concern regarding aquatic toxicity.[80] Additional publicly available peer-reviewed studies would be helpful to confirm the safety of SLG in the environment.

5.2.3 Technical Summaries for Rhamnolipid and Sodium Lauroyl Glutamate Surfactants

Rhamnolipids and sodium lauroyl glutamate (SLG) both have the technical potential to replace fluorosurfactants in floor polish formulations. Both alternatives reduced the surface tension of water by about 50%[45,79] at a concentration[50,79] comparable to that of fluorosurfactants.[46,47,92] Rhamnolipids displayed strong wettability performance with a low contact angle on PVC.[51] SLG may have better recoatability potential than rhamnolipids based on its low $\log K_{ow}$.[80] A summary of the technical performance metrics and their overall efficiencies for fluorosurfactants, SDS, rhamnolipids, and SLG are shown in Figure 5.6.

This chapter outlined two strategies for tuning the properties of rhamnolipids and amino acid surfactants: changing the salt counterion or concentration and using gemini surfactant forms. Another strategy that may be useful for tuning surfactant properties is to mix different surfactants together. Synergistic interactions can occur in surfactant mixtures, which often lead to smaller surface tensions and critical micelle concentrations than their single component counterparts for rhamnolipids, amino acids, and gemini surfactants.[93–97] For example, surface tension drastically decreased to below 1 $mN\,m^{-1}$ when rhamnolipids were combined with a synthetic non-fluorinated surfactant.[93] Combining surfactants also tuned other surface activities such as micelle formation.[94,96] Overall, a possible strategy for overcoming technical deficits with one surfactant may be to add another surfactant.

5.2.4 Hazard Assessment Summaries for Rhamnolipid and Sodium Lauroyl Glutamate Surfactants

Rhamnolipids and SLG demonstrated low hazard concerns across most endpoints, especially compared to PFASs, as shown in Figure 5.7. Most importantly, both alternatives degraded[44,45,67,80] and did not persist in the environment.[65,66,80] Both candidates demonstrated low overall toxicological concern, with SLG possibly exhibiting a lower potential for aquatic toxicity than rhamnolipids.[73,74,99] However, both assessments were based on limited studies and modeled data; thus, additional studies on the aquatic toxicity of these surfactants are needed in order to make informed conclusions.

Rhamnolipids clearly pose minimal concern across the human health endpoints, while SLG possibly presents low concern but requires more studies. The main concerns for the production of rhamnolipids are based on the use of a bacterial host and hazardous solvents.[61] However, there are clear strategies and alternatives (*e.g.*, safety protocols,[60] green solvents[63,64]) that reduce occupational exposure risks during production, and may also be relevant in preventing skin and eye irritation during the product use stage.

5.3 Importance, Innovation, and Impact

Finding PFAS replacements through functional substitution rather than chemical substitution expands the possibilities for incorporating safer alternatives.[147] PFASs dominate floor polish formulations because they excel in leveling and wetting capabilities at low concentrations while also providing slip resistance, durability, and a glossy appearance. In the past, the chemical substitution or "drop-in" approach has resulted in the replacement of PFASs with other PFASs – often a short-chain PFAS – that may not be explicitly banned but that still carry environmental and health risks.[147–149] Approaching chemical assessment through the lens of functional substitution allows for broad thinking about alternative chemicals, materials, products, processes, and systems for achieving a particular chemical function, beyond simple drop-in chemical substitutes (that may have similar toxicity profiles as the substituted chemical) and, as a result, supports a transition to safer, functionally equivalent alternatives.[147] The Maine State Legislature strengthened this approach through legislation,[4] which requires reporting of intentionally-added PFASs in products by 2023 and enables the Maine Department of Environmental Protection (DEP) to ban PFASs in other products or product categories if safer alternatives exist and the use of PFASs in the product is non-essential. This legislation stimulated the pursuit of non-fluorinated surfactants in products, and this chapter's focus on floor polish is an example of how the DEP and other stakeholders can search for and provide guidance on specific examples of alternatives to PFASs in consumer products for industries to explore.

While removing PFASs from floor polish and using alternatives are necessary, it is important to note that there are also other potentially hazardous ingredients in many floor polish formulations such as volatile organic compounds, harmful preservatives, and styrene copolymers.[150] Using products with these chemicals could lead to additional occupational, consumer, and environmental exposure risks. Exposure to other ingredients in floor polish, such as styrene copolymers, is associated with health outcomes such as cancer, brain damage, birth defects, and asthma.[151,152] Therefore, the issue of finding safe and eco-friendly chemicals in floor polish products beyond PFAS replacements challenges the floor polish industry to consider a paradigm shift in overall formulation.

Possibilities for eliminating exposure to PFASs from their presence in floor polish extends beyond formulation changes. While floor polish makes floors more durable and visually appealing, consumers can choose flooring material that is sustainable and durable and does not require polishing. For example, Syracuse University recently installed rubber flooring that was easy to clean, visually appealing, and did not require floor polish, in a number of spaces on their campus.[153] The university's primary goal in finding an alternative flooring material was the easy maintenance, aiming to eliminate the need for stripping and waxing, thus avoiding the need to shut down busy dorms or large areas on campus for a long time while waiting for floor wax to be applied and left to dry.[153] The use of alternative flooring materials has the potential to end the need for regularly stripping and repolishing floors.

Removing PFASs from floor polish will protect individuals from future contamination of well water, as PFASs can enter the water system through multiple pathways during the manufacture, application, and disposal of floor polish.[4] Water contamination can lead to the accumulation of PFASs in soil, plants, aquatic organisms, and eventually humans, which leads to an increased risk of adverse health outcomes.[14,154,155] Eliminating PFASs from floor polish would impact large buildings with high-volume floor polish use, such as schools, nursing homes, and hospitals. This elimination would benefit sensitive and vulnerable populations (*e.g.*, children and the elderly) in those settings, as well as floor waxing staff and floor polish producers. One study estimated PFAS exposure (specifically perfluoro-2-methoxyacetic acid, perfluorobutanoic acid, perfluorohexanoic acid, perfluoroheptanoic acid, and perfluorooctane sulfonic acid) during floor stripping and waxing and found measurements one to two orders of magnitude beyond ambient concentrations.[7] Therefore, eliminating PFAS exposure, rather than solely reducing the exposure using safety protocols such as personal protective equipment, will provide immense benefits to floor waxing workers.[7]

Currently, the Maine DEP is collecting industry reports of PFAS applications to share with the state by the beginning of 2023, according to an email update from the Defend Our Health partner.[156] Once these reports are examined, the replacement of PFASs in additional categories of consumer products may be considered, with floor polish being used as an example product category. The recommendations for alternative surfactants in floor polish in this chapter served as an example of how functional substitution can lead to solutions beyond the PFAS class.

5.4 Application

5.4.1 Lessons Learned and Recommendations

Due to the diverse composition of floor polish formulations and the limited data on PFASs in those products, real-world predictions for technical performances and health and environmental impacts were challenging to estimate and may differ across different formulations. Studies on surfactants provided different metrics and conditions for performance or impacts, so it was crucial to translate results across studies, establish limitations for comparison, or find similar structures for comparison in the absence of data. A major limitation on the predicted performance of rhamnolipid and SLG surfactants in floor polish was the lack of knowledge about the expected reactions with other compounds in the floor polish formulation. The recommendations provided in this chapter will serve to enhance future assessments of the viability of rhamnolipids and SLG in floor polish formulations and will give a more comprehensive understanding of their health and environmental impacts.

It will be helpful to further study the wettability potential of both rhamnolipids and SLG on a larger variety of surfaces that are relevant for floor polish applications. Although this work led to the prediction that SLG may have better recoatability potential than rhamnolipids, the best way to evaluate recoatability will be to test the performance of these alternative surfactants when mixed into a floor polish solution. Testing mixtures of non-fluorinated surfactants will also yield insight into whether they can synergistically enhance the surfactant performance.

More studies on the aquatic toxicity of rhamnolipids and SLG surfactants are needed since the assessments were based on limited studies and modeled data.[73,74,99] This information will reconcile whether rhamnolipids truly demonstrate aquatic toxicity or not. More information is needed in order to better estimate the concentrations required for floor polish use, and these values will be helpful in establishing safety procedures and personal protective equipment for the production and application of floor polish containing these non-fluorinated surfactants. Additional research on carcinogenicity, reproductive and developmental toxicity, endocrine disruption, and neurotoxicity will help fill data gaps and increase confidence in hazard score assignments. Furthermore, studies that provide more detailed information on the exposure routes will assist in better evaluating the hazards for surfactants and implementing effective safety measures.

5.4.2 Outlook

Although both rhamnolipids and SLG were technically comparable, rhamnolipids are commercially used in a larger variety of industries than amino acid surfactants, so they offer a good initial starting point for investigating PFAS replacement in floor polish. Rhamnolipids had a larger number of studies evaluating their surfactant properties and hazard outcomes, which ultimately led to this recommendation. While amino acid surfactants are growing in commercial applications, finding comprehensive studies and information that were specific to an amino acid surfactant was challenging given the vast diversity of possible amino acid surfactants. Figure 5.8 summarizes the general advantages (blue checkmarks) and disadvantages (red/pink x-marks) of the alternative candidates discussed in this work.

Beyond this initial exploration of these alternative surfactants, the performance of rhamnolipids mixed into floor polish formulations needs to be tested. First, the necessary concentration for rhamnolipids needs to be determined. Ideally, a low concentration reduces production costs and minimizes modifications of the final floor polish properties.[10] Second, laboratory tests need to evaluate the effectiveness of floor polish solutions with rhamnolipids on different floor materials. A longer-term direction could entail pursuing one of the surfactant optimization strategies outlined earlier (*i.e.*, salt counterion, gemini surfactants, or mixing surfactant systems together). Ultimately, the pursuit of PFAS replacements through

	Technical	Human Health	Environmental
Rhamnolipids	✓ Strong leveling and wetting potential ✓ More available studies ✓ Currently used in broad commercial applications	✓ Low concern across human health endpoints X Data gaps in literature	✓ More studies on environmental outcomes demonstrating safety ✓ Biodegrades and helps degrade other pollutants X Possible aquatic toxicity
Sodium lauroyl glutamate (SLG)	✓ Strong potential for leveling and recoating X Not as widely used in industry	✓ On EPA's List of Safer Chemical Ingredients X Data gaps in literature	✓ Appears safe and is able to biodegrade X Insufficient studies

Figure 5.8 Summary of the technical, human health, and environmental assessment of the two proposed solutions, rhamnolipids and sodium lauroyl glutamate, with their respective advantages (blue checkmarks) and disadvantages (red/pink x-marks).

functional substitution and paradigm shifts in floor polish formulations and flooring material will be motivated by legislation, regulations, and informed consumers.

References

1. Maine Discovers New Toxic PFAS Contamination in Drinking Water, https://defendourhealth.org/news/maine-discovers-new-toxic-pfas-contamination-in-drinking-water/ (accessed 15 December 2021).
2. Stoneridge Farm Background Documents for PFOS Contamination Concern, https://www.nebiosolids.org/stoneridge-farm-background-docs (accessed 30 September 2022).
3. Despite Toxic 'Forever Chemicals' in Trenton School's Drinking Water, State Failing to Require Action, https://defendourhealth.org/news/despite-toxic-forever-chemicals-in-trenton-schools-drinking-water-state-failing-to-require-action/ (accessed 30 September 2022).
4. L. Gramlich, LD 1503, HP 1113, An Act To Stop Perfluoroalkyl and Polyfluoroalkyl Substances Pollution, 130th Maine Legislature, First Special Session, 2021.
5. Maine PFAS Task Force, Managing PFAS in Maine – Final Report from the Maine PFAS Task Force, 2020.
6. C. Savaglio, Zoom Conversation with Carmine, 2021.
7. J. Zhou, K. Baumann, N. Chang, G. Morrison, W. Bodnar, Z. Zhang, J. M. Atkin, J. D. Surratt and B. J. Turpin, *Atmos. Environ.*, 2022, **268**, 118845.

8. D. E. P. Agency, Occurrence in consumer products – Survey of Chemical Substances in Consumer Products, https://www2.mst.dk/udgiv/publications/2008/978-87-7052-845-0/html/kap04_eng.htm (accessed 13 July 2022).
9. E. Pemberton, Poly- and perfluoroalkyl substances (PFAS): sources, pathways and environmental data, Environment Agency, Bristol, UK, 2021.
10. C. Savaglio, Fundamentals of Floor Care with Carmine Savaglio through Zoom Presentation, 2021.
11. Chemours Fluorochemicals, Capstone™ Fluorosurfactants: For high value-in-use applications that require maximum performance, https://www.chemours.com/en/-/media/files/capstone/capstone-surfactants-brochure.pdf?rev=7afbeb3f1b984d3c99e225bd2c9edc84 (accessed 1 November 2021).
12. Glycine, *N*-ethyl-*N*-((1,1,2,2,3,3,4,4,5,5,6,6,7,7,8,8,8-heptadecafluorooctyl)sulfonyl)-, https://pharosproject.net/chemicals/2011683 (accessed 29 December 2022).
13. E. W. Group, EWG's Tap Water Database, https://www.ewg.org/tapwater/contaminant.php?contamcode=E310?contamcode=E310 (accessed 29 December 2022).
14. L. P. Burkhard, *Environ. Toxicol. Chem.*, 2021, **40**, 1530–1543.
15. US EPA and OCSPP, KABAM Version 1.0 User's Guide and Technical Documentation – Appendix F – Description of Equations Used to Calculate the BCF, BAF, BMF, and BSAF Values, https://www.epa.gov/pesticide-science-and-assessing-pesticide-risks/kabam-version-10-users-guide-and-technical-3 (accessed 27 November 2021).
16. Perfluorooctanesulfonic Acid (PFOS, C-8), https://pharosproject.net/chemicals/2011510#hazards-panel (accessed 17 December 2021).
17. Z. Zeng, B. Song, R. Xiao, G. Zeng, J. Gong, M. Chen, P. Xu, P. Zhang, M. Shen and H. Yi, *Environ. Int.*, 2019, **126**, 598–610.
18. Perfluorobutanesulfonic acid, https://pharosproject.net/chemicals/2012913#hazards-panel (accessed 17 November 2021).
19. Evonik Products, DYNOL™ 800, https://www.productcenter.coating-additives.com/pdf/daten/engl/DYNOL_800.pdf (accessed 14 July 2022).
20. Evonik Products, DYNOL™ 810, https://www.productcenter.coating-additives.com/pdf/daten/engl/DYNOL_810.pdf (accessed 14 July 2022).
21. Green Surfactants For Sustainable Cleaning, https://www.mahachem.com/green-surfactants-for-sustainable-cleaning/ (accessed 15 July 2022).
22. L. Yan, Alkyl polyglucoside applications in industry, https://www.irocoatingadditive.com/alkyl-polyglucoside-applications-industry/ (accessed 14 December 2021).
23. Dow Consumer & Industrial Solutions, TRITON™ CG-425 Alkyl Polyglucoside Surfactant, https://www.dow.com/content/dam/dcc/documents/en-us/mark-prod-info/805/805-001/805-00047-01-triton-cg425-alkyl-polyglucoside-surfactant-product-guide.pdf?iframe=true (accessed 16 December 2021).

24. Y. Wang and 3M, Aqueous Floor Wax Composition Patent Application, https://patentimages.storage.googleapis.com/23/94/ad/badc592429cd16/WO2008134180A1.pdf (accessed 14 July 2022).
25. Defend Our Health|Solutions for a Toxic Free Tomorrow, https://defendourhealth.org/ (accessed 19 September 2021).
26. S. Greene and C. Neuschler, Minnesota's PFAS Blueprint, Minnesota Pollution Control Agency, Saint Paul, MN, 2021.
27. A. Jarray, H. Wijshoff, J. A. Luiken and W. K. den Otter, *Soft Matter*, 2020, **16**, 4299–4310.
28. S. Laurén, What is wettability?, https://www.biolinscientific.com/blog/what-is-wettability (accessed 10 November 2021).
29. Water Science School, Surface Tension and Water, https://www.usgs.gov/special-topics/water-science-school/science/surface-tension-and-water (accessed 3 August 2022).
30. T. Karbowiak, F. Debeaufort and A. Voilley, *Crit. Rev. Food Sci. Nutr.*, 2006, **46**, 391–407.
31. A. Czajka, G. Hazell and J. Eastoe, *Langmuir*, 2015, **31**, 8205–8217.
32. Critical micelle concentration (CMC) and surfactant concentration, https://www.kruss-scientific.com/en-US/know-how/glossary/critical-micelle-concentration-cmc-and-surfactant-concentration (accessed 10 November 2021).
33. H. Cumming and C. Rücker, *ACS Omega*, 2017, **2**, 6244–6249.
34. Stockholm Convention on Persistant Organic Pollutants (POPs) – Annex D, Secretariat of the Stockholm Convention, 2019.
35. I. T. Cousins, J. C. DeWitt, J. Glüge, G. Goldenman, D. Herzke, R. Lohmann, C. A. Ng, M. Scheringer and Z. Wang, *Environ. Sci.: Processes Impacts*, 2020, **22**, 2307–2312.
36. J. Glüge, R. London, I. T. Cousins, J. DeWitt, G. Goldenman, D. Herzke, R. Lohmann, M. Miller, C. A. Ng, S. Patton, X. Trier, Z. Wang and M. Scheringer, *Environ. Sci. Technol.*, 2022, **56**, 6232–6242.
37. GreenScreen® Method, https://www.greenscreenchemicals.org/learn/full-greenscreen-method (accessed 3 August 2022).
38. O. P. Ward, in *Biosurfactants*, ed. R. Sen, Springer Science+Business Media, Landes Bioscience, New York, NY, Austin, TX, 1st edn, 2010, pp. 65–74.
39. G. Ocampo, *J. Bacteriol. Mycol.*, 2016, **2**, 95–96.
40. K. Kamaljeet, S. Randhawa and P. K. S. M. Rahman, *Front. Microbiol.*, 2014, **5**, 454.
41. G. Rawat, A. Dhasmana and V. Kumar, *Environ. Sustainability*, 2020, **3**, 353–369.
42. Y. Ruiz-Morales and A. Romero-Martínez, *J. Phys. Chem. B*, 2018, **122**, 3931–3943.
43. M. M. Müller, J. H. Kügler, M. Henkel, M. Gerlitzki, B. Hörmann, M. Pöhnlein, C. Syldatk and R. Hausmann, *J. Biotechnol.*, 2012, **162**, 366–380.

44. Top 20 innovative bio-based products, Task 3 of "Study on Support to R&I Policy in the Area of Bio-based Products and Services", Released December 2018, DOI: 10.2777/85805, ISBN 978-92-76-03420-9.
45. J. Liepins, K. Balina, R. Soloha, I. Berzina, L. K. Lukasa and E. Dace, *Fermentation*, 2021, **7**, 136.
46. Chemours, Capstone FS-60 Fluorosurfactant, https://www.chemours.com/es/-/media/files/capstone/capstone-fs-60-technical-info.pdf?rev=f8fcd51e4a83496ebdadc09261202a4e&hash=F03BB0E41BFE588D13313FB7271A9B6B (accessed 17 November 2021).
47. Chemours, Capstone FS-65 Fluorosurfactant, https://www.chemours.com/en/-/media/files/capstone/capstone-fs-65-technical-info.pdf?rev=9c6f5d62555b44ee868e173f8e677d39&hash=BD3E31321F1D3A10F7938C268C901357 (accessed 17 November 2021).
48. C. Hill, A. Czajka, G. Hazell, I. Grillo, S. E. Rogers, M. W. A. Skoda, N. Joslin, J. Payne and J. Eastoe, *J. Colloid Interface Sci.*, 2018, **530**, 686–694.
49. S. Mall, G. Buckton and D. A. Rawlins, *J. Pharm. Sci.*, 1996, **85**, 75–78.
50. L.-M. Wu, L. Lai, Q. Lu, P. Mei, Y.-Q. Wang, L. Cheng and Y. Liu, *Colloids Surf., B*, 2019, **181**, 593–601.
51. S. Costa, S. R. de Souza, M. Nitschke, S. M. M. Franchetti, M. Jafelicci Jr, R. B. Lovaglio and J. Contiero, *J. Surfactants Deterg.*, 2009, **12**, 125–130.
52. 2-(*N*-Ethylperfluorooctanesulfonamido)acetic acid Properties, https://comptox.epa.gov/dashboard/chemical/properties/DTXSID5062760 (accessed 29 December 2022).
53. US EPA Comptox Dashboard, Rhamnolipid RL2, https://comptox.epa.gov/dashboard/chemical/details/DTXSID80884062 (accessed 27 September 2022).
54. J. Costanza, M. Arshadi, L. M. Abriola and K. D. Pennell, *Environ. Sci. Technol. Lett.*, 2019, **6**, 487–491.
55. R. Krustev, H. J. Müller and J. L. Toca-Herrera, *Colloid Polym. Sci.*, 1998, **276**, 518–523.
56. S. S. Helvaci, S. Peker and G. Ozdemir, *Colloids Surf., B*, 2004, **35**, 225–233.
57. K. Patowary, M. Das, R. Patowary, M. C. Kalita and S. Deka, *J. Surfactants Deterg.*, 2019, **22**, 373–384.
58. S. Semkova, G. Antov, I. Iliev, I. Tsoneva, P. Lefterov, N. Christova, L. Nacheva, I. Stoineva, L. Kabaivanova, G. Staneva and B. Nikolova, *Separations*, 2021, **8**, 92.
59. S. P. Diggle and M. Whiteley, *Microbiology*, 2020, **166**, 30–33.
60. Pseudomonas Infection|Michigan Medicine, https://www.uofmhealth.org/health-library/abj6977 (accessed 15 December 2021).
61. H. Chong and Q. Li, *Microb. Cell Fact.*, 2017, **16**, 137.
62. L. Dobler, H. C. Ferraz, L. V. Araujo de Castilho, L. S. Sangenito, I. P. Pasqualino, A. L. Souza dos Santos, B. C. Neves, R. R. Oliveira, D. M. Guimarães Freire and R. V. Almeida, *Chemosphere*, 2020, **252**, 126349.

63. P. Eslami, H. Hajfarajollah and S. Bazsefidpar, *RSC Adv.*, 2020, **10**, 34014–34032.
64. F. P. Byrne, S. Jin, G. Paggiola, T. H. M. Petchey, J. H. Clark, T. J. Farmer, A. J. Hunt, C. Robert McElroy and J. Sherwood, *Sustainable Chem. Processes*, 2016, **4**, 7.
65. Rhamnolipids: fermentation products of glucose with Pseudomonas bacteria, https://echa.europa.eu/sl/registration-dossier/-/registered-dossier/20930/5/3/2 (accessed 27 September 2022).
66. US EPA Office of Pesticide Programs, *Biopesticides Registration Action Document: Rhamnolipid Biosurfactant*, US Environmental Protection Agency – Office of Pesticides Program, 2004.
67. P. K. Mohan, G. Nakhla and E. K. Yanful, *Water Res.*, 2006, **40**, 533–540.
68. Ł. Chrzanowski, M. Dziadas, Ł. Ławniczak, P. Cyplik, W. Białas, A. Szulc, P. Lisiecki and H. Jeleń, *Bioresour. Technol.*, 2012, **111**, 328–335.
69. B. Dahrazma and C. N. Mulligan, *Chemosphere*, 2007, **69**, 705–711.
70. K. A. Vu, K. Tawfiq and G. Chen, *Water, Air, Soil Pollut.*, 2015, **226**, 256.
71. Z. Wei, J. J. Wang, Y. Meng, J. Li, L. A. Gaston, L. M. Fultz and R. D. DeLaune, *Chemosphere*, 2020, **253**, 126617.
72. G. Hodges, C. Eadsforth, B. Bossuyt, A. Bouvy, M.-H. Enrici, M. Geurts, M. Kotthoff, E. Michie, D. Miller, J. Müller, G. Oetter, J. Roberts, D. Schowanek, P. Sun and J. Venzmer, *Environ. Sci. Eur.*, 2019, **31**, 1–18.
73. 2-O-Rhamnopyranosyl-rhamnopyranosyl-3-hydroxyldecanoyl-3-hydroxydecanoate, https://pharosproject.net/chemicals/2194944#hazards-panel (accessed 16 December 2021).
74. S. Johann, T.-B. Seiler, T. Tiso, K. Bluhm, L. M. Blank and H. Hollert, *Sci. Total Environ.*, 2016, **548–549**, 155–163.
75. M. C. Morán, A. Pinazo, L. Pérez, P. Clapés, M. Angelet, M. T. García, M. P. Vinardell and M. R. Infante, *Green Chem.*, 2004, **6**, 233–240.
76. L. Pinheiro and C. Faustino, in *Application and Characterization of Surfactants*, ed. R. Najjar, IntechOpen Limited, London, United Kingdom, 1st edn, 2017, DOI: 10.5772/67977.
77. Understanding Surfactants, https://www.ipcol.com/blog/an-easy-guide-to-understanding-surfactants/ (accessed 15 December 2021).
78. E. Scott, F. Peter and J. Sanders, *Appl. Microbiol. Biotechnol.*, 2007, **75**, 751–762.
79. Q. Wang, Z. Song, F. Han, B. Xu and B. Xu, *Colloids Surf., A*, 2022, **640**, 128474.
80. ECHA, Sodium hydrogen *N*-(1-oxododecyl)-L-glutamate, Registration Dossier – ECHA, https://echa.europa.eu/de/registration-dossier/-/registered-dossier/20719/5/4/1 (accessed 23 November 2022).
81. L. Pérez, A. Pinazo, R. Pons and M. Infante, *Adv. Colloid Interface Sci.*, 2014, **205**, 134–155.
82. A. Pinazo, R. Pons, L. Pérez and M. R. Infante, *Ind. Eng. Chem. Res.*, 2011, **50**, 4805–4817.
83. N. Pal, H. Hoteit and A. Mandal, *J. Mol. Liq.*, 2021, **339**, 116811.

84. T. Yoshimura and K. Esumi, *J. Colloid Interface Sci.*, 2004, **276**, 231–238.
85. L. R. Dix, *J. Colloid Interface Sci.*, 2001, **238**, 447–448.
86. M. Akram, F. Ansari, I. A. Bhat and Kabir-ud-Din, *J. Mol. Liq.*, 2019, **276**, 519–528.
87. US EPA, Safer Chemical Ingredients List, https://sor.epa.gov/sor_internet/registry/substreg/searchandretrieve/substancesearch/search.do?synId=1241605&displaySynonym= (accessed 16 December 2021).
88. ECHA, Sodium hydrogen *N*-(1-oxododecyl)-L-glutamate, Substance Information Card – ECHA, https://echa.europa.eu/substance-information/-/substanceinfo/100.045.402 (accessed 23 November 2022).
89. Sodium lauroyl glutamate, https://pharosproject.net/chemicals/2077300#hazards-panel (accessed 13 December 2021).
90. C. L. Burnett, B. Heldreth, W. F. Bergfeld, D. V. Belsito, R. A. Hill, C. D. Klaassen, D. C. Liebler, J. G. Marks, R. C. Shank, T. J. Slaga, P. W. Snyder and F. A. Andersen, *Int. J. Toxicol.*, 2017, **36**, 17S–56S.
91. K. Parsi, *Phlebology*, 2015, **30**, 306–315.
92. A. Banerjee and Y. Liu, *Langmuir*, 2021, **37**, 8937–8944.
93. T. T. Nguyen, N. H. Youssef, M. J. McInerney and D. A. Sabatini, *Water Res.*, 2008, **42**, 1735–1743.
94. M. Chen, C. Dong, J. Penfold, R. K. Thomas, T. J. P. Smyth, A. Perfumo, R. Marchant, I. M. Banat, P. Stevenson, A. Parry, I. Tucker and I. Grillo, *Langmuir*, 2013, **29**, 3912–3923.
95. W. Zhang, Z. Gao, H. Zhu and Q. Zhang, *J. Dispersion Sci. Technol.*, 2021, **0**, 1–13.
96. X. Ji, M. Tian and Y. Wang, *Langmuir*, 2016, **32**, 972–981.
97. K. S. Sharma, C. Rodgers, R. M. Palepu and A. K. Rakshit, *J. Colloid Interface Sci.*, 2003, **268**, 482–488.
98. C. Guo, C. K. Chen, W. Zhou, J. Meng and J. Chou, Novel Application of Fluorosurfactants in PCI Magazine, https://www.pcimag.com/articles/92574-novel-application-of-fluorosurfactants-in (accessed 6 July 2022).
99. L-Glutamic acid, *N*-(1-oxododecyl)-, sodium salt (1:1) – Chemical Details, https://comptox.epa.gov/dashboard/chemical/details/DTXSID90885464 (accessed 27 September 2022).
100. J. J. Shearer, C. L. Callahan, A. M. Calafat, W.-Y. Huang, R. R. Jones, V. S. Sabbisetti, N. D. Freedman, J. N. Sampson, D. T. Silverman, M. P. Purdue and J. N. Hofmann, *J. Natl. Cancer Inst.*, 2021, **113**, 580–587.
101. B. A. Cohn, M. A. La Merrill, N. Y. Krigbaum, M. Wang, J.-S. Park, M. Petreas, G. Yeh, R. C. Hovey, L. Zimmermann and P. M. Cirillo, *Reprod. Toxicol.*, 2020, **92**, 112–119.
102. Environmental Protection Agency and D. of Toxic Substances Control, Product-Chemical Profile for Food Packaging with PFASs: Discussion Draft, https://dtsc.ca.gov/wp-content/uploads/sites/31/2020/07/Draft-Profile_PFASs-in-Food-Packaging_FINAL_ADA.pdf (accessed 27 September 2022).
103. T. M. O'Brien, R. M. Carlson, P. J. Oliveira and K. B. Wallace, *Chem. Res. Toxicol.*, 2006, **19**, 1305–1312.

104. E. V. Preston, T. F. Webster, E. Oken, B. Claus Henn, M. D. McClean, S. L. Rifas-Shiman, E. N. Pearce, L. E. Braverman, A. M. Calafat, X. Ye and S. K. Sagiv, *Environ. Health Perspect.*, 2018, **126**, 027013.
105. A. Cardenas, D. R. Gold, R. Hauser, K. P. Kleinman, M.-F. Hivert, A. M. Calafat, X. Ye, T. F. Webster, E. S. Horton and E. Oken, *Environ. Health Perspect.*, 2017, **125**(10), 107001.
106. C. Zhang, R. Sundaram, J. Maisog, A. M. Calafat, D. B. Barr and G. M. Buck Louis, *Fertil. Steril.*, 2015, **103**, 184–189.
107. C. Xiao, P. Grandjean, D. Valvi, F. Nielsen, T. K. Jensen, P. Weihe and Y. Oulhote, *J. Clin. Endocrinol. Metab.*, 2020, **105**, 735–745.
108. E. V. Preston, S. L. Rifas-Shiman, M.-F. Hivert, A. R. Zota, S. K. Sagiv, A. M. Calafat, E. Oken and T. James-Todd, *J. Clin. Endocrinol. Metab.*, 2020, **105**, e2864–e2876.
109. M. Berk, L. J. Williams, A. C. Andreazza, J. A. Pasco, S. Dodd, F. N. Jacka, S. Moylan, E. J. Reiner and P. V. S. Magalhaes, *BMJ Open*, 2014, **4**(7), e005142.
110. K. Lyall, V. M. Yau, R. Hansen, M. Kharrazi, C. K. Yoshida, A. M. Calafat, G. Windham and L. A. Croen, *Environ. Health Perspect.*, 2018, **126**, 017001.
111. J. Oh, D. H. Bennett, A. M. Calafat, D. Tancredi, D. L. Roa, R. J. Schmidt, I. Hertz-Picciotto and H.-M. Shin, *Environ. Int.*, 2021, **147**, 106328.
112. H.-M. Shin, D. H. Bennett, A. M. Calafat, D. Tancredi and I. Hertz-Picciotto, *Environ. Res.*, 2020, **186**, 109514.
113. I. Shiue, *Environ. Sci. Pollut. Res.*, 2015, **22**, 6381–6386.
114. I. Shiue, *Environ. Sci. Pollut. Res.*, 2015, **22**, 17095–17103.
115. I. Shiue, *Environ. Sci. Pollut. Res. Int.*, 2017, **24**, 3108–3116.
116. K. E. Reed, Exhibit 1737: Letter from K. Reed to Document Processing Center (7407) re: TSCA Section 8(e) Supplemental Submissions: Docket Nos. 8EHQ-1180-373 and 8EHQ-1180-374 [Internet], The Office of Minnesota Attorney General, 2000, https://www.ag.state.mn.us/Office/Cases/3M/StatesExhibits.asp.
117. C. Schilliger-Musset, *Inclusion of substances of very high concern in the Candidate List for eventual inclusion in Annex XIV (Decision of the European Chemicals Agency)*, 2020.
118. H.-W. Lin, H.-X. Feng, L. Chen, X.-J. Yuan and Z. Tan, *Nagoya J. Med. Sci.*, 2020, **82**, 323–333.
119. R. A. Brase, E. J. Mullin and D. C. Spink, *Int. J. Mol. Sci.*, 2021, **22**, 995.
120. K. T. Eriksen, O. Raaschou-Nielsen, M. Sørensen, M. Roursgaard, S. Loft and P. Møller, *Mutat. Res., Genet. Toxicol. Environ. Mutagen.*, 2010, **700**, 39–43.
121. M. Blanc, J. Rüegg, N. Scherbak and S. H. Keiter, *Aquat. Toxicol.*, 2019, **215**, 105272.
122. S. Liu, R. Yang, N. Yin and F. Faiola, *J. Environ. Sci.*, 2020, **88**, 187–199.
123. US EPA and ORD, Learn about the Human Health Toxicity Assessment for PFBS, https://www.epa.gov/chemical-research/learn-about-human-health-toxicity-assessment-pfbs (accessed 27 September 2022).

124. X. Feng, X. Cao, S. Zhao, X. Wang, X. Hua, L. Chen and L. Chen, *Toxicol. Sci*, 2017, **155**, 409–419.
125. L. Chen, M. M. P. Tsui, C. Hu, T. Wan, D. W. T. Au, J. C. W. Lam, P. K. S. Lam and B. Zhou, *Environ. Sci. Technol.*, 2019, **53**, 12018–12025.
126. K. M. Annunziato, M. Marin, W. Liang, S. M. Conlin, W. Qi, J. Doherty, J. Lee, J. M. Clark, Y. Park and A. R. Timme-Laragy, *Chemosphere*, 2022, **287**, 132121.
127. L. Tang, M. Liu, S. Song, C. Hu, P. K. S. Lam, J. C. W. Lam and L. Chen, *Aquat. Toxicol.*, 2020, **222**, 105466.
128. O. Wasel, K. M. Thompson, Y. Gao, A. E. Godfrey, J. Gao, C. T. Mahapatra, L. S. Lee, M. S. Sepúlveda and J. L. Freeman, *J. Toxicol. Environ. Health, Part A*, 2021, **84**, 125–136.
129. L. Tang, S. Song, C. Hu, M. Liu, P. K. S. Lam, B. Zhou, J. C. W. Lam and L. Chen, *Chemosphere*, 2020, **256**, 127169.
130. Agency for Toxic Substances and Disease Registry, M. Buser, H. R. Pohl, P. Ruiz, F. Scinicariello, S. Chou, H. Abadin, L. Ingerman, L. Barber, H. Carlson-Lynch, M. Citra, G. L. Diamond, J. Klotzbach, F. T. Llados, D. J. Plewak and SRC, Inc., *Toxicological profile for Perfluoroalkyls*, Agency for Toxic Substances and Disease Registry, 2021.
131. S. Gaballah, A. Swank, J. R. Sobus, X. M. Howey, J. Schmid, T. Catron, J. McCord, E. Hines, M. Strynar and T. Tal, *Environ. Health Perspect.*, 2020, **128**, 047005.
132. M. Liu, S. Song, C. Hu, L. Tang, J. C. W. Lam, P. K. S. Lam and L. Chen, *Environ. Pollut.*, 2020, **265**, 114832.
133. T. A. Slotkin, E. A. MacKillop, R. L. Melnick, K. A. Thayer and F. J. Seidler, *Environ. Health Perspect.*, 2008, **116**, 716–722.
134. Q. Zhang, W. Liu, Q. Niu, Y. Wang, H. Zhao, H. Zhang, J. Song, S. Tsuda and N. Saito, *Toxicol. Res.*, 2016, **5**, 539–546.
135. P. H. Lieder, S.-C. Chang, R. G. York and J. L. Butenhoff, *Toxicology*, 2009, **255**, 45–52.
136. H. Huang, K. Yu, X. Zeng, Q. Chen, Q. Liu, Y. Zhao, J. Zhang, X. Zhang and L. Huang, *Environ. Res.*, 2020, **191**, 110156.
137. X.-D. Qin, Z. (Min) Qian, S. C. Dharmage, J. Perret, S. D. Geiger, S. E. Rigdon, S. Howard, X.-W. Zeng, L.-W. Hu, B.-Y. Yang, Y. Zhou, M. Li, S.-L. Xu, W.-W. Bao, Y.-Z. Zhang, P. Yuan, J. Wang, C. Zhang, Y.-P. Tian, M. Nian, X. Xiao, W. Chen, Y. L. Lee and G.-H. Dong, *Environ. Res.*, 2017, **155**, 15–21.
138. E. Emerce and Ö. Çetin, *Toxicol. Ind. Health*, 2018, **34**, 884–890.
139. S. Fuentes, P. Vicens, M. T. Colomina and J. L. Domingo, *Toxicology*, 2007, **242**, 123–129.
140. N. Johansson, A. Fredriksson and P. Eriksson, *Neurotoxicology*, 2008, **29**, 160–169.
141. C. Lau, K. Anitole, C. Hodes, D. Lai, A. Pfahles-Hutchens and J. Seed, *Toxicol. Sci*, 2007, **99**, 366–394.
142. J. B. Sørli, M. Låg, L. Ekeren, J. Perez-Gil, L. S. Haug, E. Da Silva, M. N. Matrod, K. B. Gützkow and B. Lindeman, *Toxicol. In Vitro*, 2020, **62**, 104656.

143. P. B. Larsen and E. Giovalli, *Perfluoroalkylated substances: PFOA, PFOS and PFOSA*, The Danish Environmental Protection Agency, Copenhagen, Denmark, 2015.
144. T. Stipcevic, C. P. Knight and T. E. Kippin, *Acta Neurol. Belg.*, 2013, **113**, 501–506.
145. A. Kyadarkunte, M. Patole and V. Pokharkar, *Cosmetics*, 2014, **1**, 159–170.
146. K. Mansouri, A. Abdelaziz, A. Rybacka, A. Roncaglioni, A. Tropsha, A. Varnek, A. Zakharov, A. Worth, A. M. Richard, C. M. Grulke, D. Trisciuzzi, D. Fourches, D. Horvath, E. Benfenati, E. Muratov, E. B. Wedebye, F. Grisoni, G. F. Mangiatordi, G. M. Incisivo, H. Hong, H. W. Ng, I. V. Tetko, I. Balabin, J. Kancherla, J. Shen, J. Burton, M. Nicklaus, M. Cassotti, N. G. Nikolov, O. Nicolotti, P. L. Andersson, Q. Zang, R. Politi, R. D. Beger, R. Todeschini, R. Huang, S. Farag, S. A. Rosenberg, S. Slavov, X. Hu and R. S. Judson, *Environ. Health Perspect.*, 2016, **124**, 1023–1033.
147. J. A. Tickner, J. N. Schifano, A. Blake, C. Rudisill and M. J. Mulvihill, *Environ. Sci. Technol.*, 2015, **49**, 742–749.
148. M. Kotthoff and M. Bücking, *Front. Chem.*, 2018, **6**, 1–6.
149. C. F. Kwiatkowski, D. Q. Andrews, L. S. Birnbaum, T. A. Bruton, J. C. DeWitt, D. R. U. Knappe, M. V. Maffini, M. F. Miller, K. E. Pelch, A. Reade, A. Soehl, X. Trier, M. Venier, C. C. Wagner, Z. Wang and A. Blum, *Environ. Sci. Technol. Lett.*, 2020, **7**, 532–543.
150. S. Yarmey, A. Paiva, A. Liu, J. Baran and 3M, Use of a Silicone Surfactant in Polishing Compositions Patent Application, https://patentimages.storage.googleapis.com/bb/81/70/8497685214805c/CA2437534C.pdf (accessed 14 December 2021).
151. Maleic anhydride, https://pharosproject.net/chemicals/2008993 (accessed 30 September 2022).
152. Styrene, https://pharosproject.net/chemicals/2010381 (accessed 30 September 2022).
153. Winter's Worst No Match for Premium Rubber Flooring at Syracuse University, https://spaces4learning.com/articles/2017/07/28/rubber-flooring.aspx (accessed 14 December 2021).
154. US EPA, Reducing PFAS in Drinking Water with Treatment Technologies, https://www.epa.gov/sciencematters/reducing-pfas-drinking-water-treatment-technologies (accessed 15 December 2021).
155. A. K. Venkatesan and R. U. Halden, *J. Hazard. Mater.*, 2013, **252–253**, 413–418.
156. P. MacRoy, *Defend Our Health Email Update*, 2022.

CHAPTER 6

Strategies for PFAS Removal During Carpet Recycling

A. MARUZZO*[a] AND S. A. BĂLAN[a,b]

[a] University of California, Berkeley, School of Public Health, Berkeley, CA, 94704, USA; [b] California Department of Toxic Substances Control, Sacramento, CA, 95812, USA
*Email: aaron_maruzzo@berkeley.edu

6.1 Challenge Overview

Since the early 1950s, the treatment of carpet fibers with perfluoroalkyl and polyfluoroalkyl substances (PFASs) has contributed to the longevity of the carpets.[1] Carpets are made of four main layers: the face fiber, primary backing, secondary backing, and carpet padding. Face fibers are typically made from synthetic materials such as nylon (*e.g.*, nylon 6 and nylon 6,6), polyethylene terephthalate (PET), and polypropylene, or from natural materials such as wool.[2,3] Most of these materials readily absorb water and oil.[2,3] As a result, the carpet industry has relied on PFAS-based treatments to impart stain repellence.[3] These treatments lower the critical tension at the carpet surface, reduce adhesion, and limit aqueous or oily substances from absorbing onto the fiber surface.[1] PFASs can be added during the manufacturing of the carpet fibers, during the manufacturing of the carpet itself, after carpet manufacture at a separate finishing facility or in stores, or by consumers using aftermarket treatments.[3] In 2011, the European Commission estimated that old carpets contained 75 parts per million (ppm) PFOS and that in-use carpets contained $\sim$1.64 million kg PFOS.[4] The total amount of PFASs in carpet in California in 2017 was estimated to be between 4.6 and 63 tonnes.[5]

Green Chemistry Series No. 81
Toward a PFAS-free Future: Safer Alternatives to Forever Chemicals
Edited by Simona A. Bălan, Thomas A. Bruton and Kimberly G. Hazard

Published by the Royal Society of Chemistry, www.rsc.org

In California, carpets account for approximately 3% by volume of the landfill waste stream.[3] In 2021, 243 million pounds of postconsumer carpets were discarded in California.[6] Of these carpets, 163 million pounds (67%) were landfilled and 67.6 million pounds (28%) were recycled.[6] Efforts to reduce carpet waste and add value to postconsumer carpets have given rise to increased recycling initiatives. For example, in 2017, the California Carpet Stewardship Program aimed to increase the rate of carpet recycling from 11% to 24%.[7] The face fiber and backing are the most recycled components of carpets, making up 98% of the recycled output in California, while the remaining carpet components are less valuable and are separated out.[8]

Most carpet recycling is carried out using a mechanical process that involves five main steps:[9]

1. Removing and collecting the carpet
2. Sorting and bailing carpets based on the face fiber type
3. Separating the face fibers from backing using a mechanical process, such as shearing, shredding, hammer milling, or rotary impact separation
4. Sorting the face fibers from the backing
5. Cleaning and re-melting the face fibers

Mechanical recycling is less expensive than chemical recycling.[10] Chemical recycling breaks down the face fibers into their original chemical components, allowing for the separation of certain toxic chemicals. So far, only nylon 6 chemical recycling has been commercialized. Mechanical recycling is likely to dominate the carpet recycling industry for the foreseeable future.

However, recycling carpet stocks as-is poses the risk of perpetuating the harms associated with PFASs (see Chapter 2). This case study highlights a semester-long investigation into possible technologies to remove PFASs from carpet fibers as part of the UC Berkeley graduate course, Greener Solutions, from August to December 2020. The student team was composed of Ned Antell (Civil and Environmental Engineering), Andrew Cullen (Public Health), and Michael Kado (Environmental Health Sciences). Course instructors (Dr Megan Schwarzman, Mr Thomas McKeag, Dr William Hart-Cooper, and Ms Kimberly Hazard), along with experts from the Department of Toxics Substances Control and carpet industry representatives, advised the student team investigating greener remediation technologies for discarded carpets. This challenge is unique in that the aim was not to discover safer substitutes to PFASs, but rather to identify processes for safer postconsumer carpet recycling.

6.2 Approach

The student team limited its scope to identifying strategies for removing fluorotelomer-based side-chain fluorinated polymers from nylon 6 and nylon 6,6 face fibers (rather than the entire carpet). Fluorotelomer-based side-chain fluorinated polymers are the most commonly used type of PFASs for carpet treatments.[3] The polymeric backbone of fluorotelomer-based side-chain fluorinated polymers is often made of acrylate or methacrylate groups. Over time, degradation of the acrylate or methacrylate groups can lead to the release of perfluoroalkyl acids (PFAAs). Thus, carpets may also contain PFAAs, but since these are monomeric impurities are not bound to any polymer backbone, any washing step during carpet recycling will likely remove the PFAAs into the aqueous phase.

Face fibers are the most recyclable component of carpets.[11] Nylon face fibers constituted over 70% of recycled carpet materials in California in 2019 and continue to have the highest market demand (Table 6.1).[8] According to industry experts, it is also expected that more PFASs are added to nylon fibers than PET fibers, supposedly due to the inherent olio- and hydrophobicity of PET.

The proposed strategies were also evaluated against five primary technical performance criteria:

1. Does the strategy remove, destroy, or do nothing to the PFAS?
2. How fast can the strategy be implemented into the current carpet recycling process?
3. How long will the necessary reaction(s) take?
4. How much energy is required?
5. How difficult will it to be to bring the technology to scale?

Each of these technical criteria were graded using a qualitative color ranking, (green is better performance and orange is worse performance) shown near the end of this chapter for all the proposed strategies (Figure 6.4). Although fiber quality is an important factor in evaluating treatment strategies, the effect of chemical treatment on fiber quality was difficult to predict and evaluating any proposed strategy at the bench scale was beyond the project scope. The student team inferred impacts on fiber quality based on the available literature, but empirical validation is still needed.

The student team considered the chemical hazards associated with each proposed approach. All chemicals involved in the reactions were included

Table 6.1 Percent by weight of recycled carpet face fibers shipped and sold in California in 2019, by fiber materials (CARE 2019).[8]

	Q1 2019 (%)	Q2 2019 (%)	Q3 2019 (%)	Q4 2019 (%)
Nylon 6	58	53	50	39
Nylon 6,6	23	32	34	35
Modular carpet tile	3	2	5	4
Polypropylene	2	2	3	2
PET	8	11	17	29

in the evaluation. Publicly available chemical hazard information was gathered from Pharos, associated Globally Harmonized System (GHS) statements, authoritative listings, and the peer-reviewed literature to score human health and environmental endpoints following the GreenScreen method. The hazard endpoints evaluated included carcinogenicity and mutagenicity (C/M); developmental and reproductive toxicity (D/R); endocrine toxicity (E); systemic toxicity and organ effects (S); neurotoxicity (N); eye and skin irritation (I); aquatic toxicity (AT); and bioaccumulation potential (B). The comparative hazard assessment for all the proposed strategies and their constituent chemicals is shown near the end of this chapter (Figure 6.5).

The alternative strategies were compared to the current baseline scenario of recycling carpets without removing the PFAS content. Any PFASs that are not removed from carpet face fibers are likely to be present in new products created from the recycled materials, such as carpets, insulation, mattresses, pillows, and automotive parts.[7] However, there is a scarcity of publicly available data about the presence and levels of PFASs in products containing recycled carpet fibers. Assuming that any PFASs present in carpets prior to recycling will remain in the fibers after recycling, the student team used PFAS estimates from virgin carpets as a conservative baseline (Figure 6.1).

6.3 Solutions

The student team identified and compared four possible strategies: base hydrolysis with granulated activated carbon (GAC), base hydrolysis with

Constituent	Median concentration ($ng\ g^{-1}$)	C/M	D/R	E	S	N	I	AT	B
6:2 FTOH	69	M[12]	M[13]	M[14]	VH[15]	DG	L[15]	H[79]	L[79]
8:2 FTOH	18.8	DG	DG	H[14]	DG	DG	DG	P[17]	VH[18]
PFHxA	10	L[19,20]	M[21,22]	DG	H[21,23]	L[21]	VH[24]	M[24]	L[25]
PFBA	8.09	DG	DG	H[14]	DG	DG	VH[26]	P[17]	DG
PFOA	6.13	H[27]	H[27]	H[14]	H[28]	DG	vH[29]	M[30,31]	VH[32]
PFPeA	4.62	DG	M[33]	H[14]	DG	DG	VH[34]	P[17]	DG
PFNA	2.7	M[35]	H[27]	H[14]	H[35]	DG	vH[35]	DG	VH[32]
PFOS	2.32	H[27]	H[36,37]	H[27]	H[38]	DG	M[39]	H[40]	VH[32]
PFHxS	1.31	DG	H[41]	H[14]	DG	DG	DG	P[17]	VH[32]

Key	High (H) / Very High (VH)	Moderate (M)	Low (L)	Potential concern (P)	Data gap (DG)
Carcinogenicity or mutagenicity (C/M), developmental or reproductive toxicity (D/R), endocrine toxicity (E), systemic toxicity and organ effects (S), neurotoxicity (N), eye and skin irritation (I), aquatic toxicity (AT), and bioaccumulation potential (B)					

Figure 6.1 Summary of health hazards associated with PFASs present in recycled carpets at levels above 1 $ng\,g^{-1}$.[12–41] Median PFAS concentrations are estimates from Wu *et al.* (2020).[42]

reverse osmosis (RO) and plasma, esterase hydrolysis, and mineralization with enzymatic laccases. Below we discuss the inspiration for each of these proposed strategies, as well as the performance in terms of technical, health, and environmental criteria.

6.3.1 Base Hydrolysis with GAC

The student team proposed using a strong base, such as sodium hydroxide or potassium hydroxide, to break (meth)acrylate linkages to release the fluorinated side-chains from carpet treatment polymers into solution. The released PFASs would then be adsorbed onto a GAC filter. This design was inspired by previous research conducted by Peaslee *et al.* (2020), who used 2M NaOH solution to extract PFASs from firefighting textiles.[43] Strong bases hydrolyze the ester group in acrylate linkages, because the carbonyl carbon is prone to nucleophilic attack by hydroxide.[44] In the case of fluorotelomer (meth)acrylates, base hydrolysis of the ester group produces fluorotelomer alcohols.[45]

Neutralizing the solution is an essential step because the high pH resulting from the strong base treatment diminishes the absorptive capacity of GAC. GAC performs better when the point of zero charge (p_{zc}) is greater than the solution pH, which facilitates electrostatic interactions.[46] If a GAC's p_{zc} is circumneutral and the solution pH is 14, the GAC functional groups deprotonate and become negatively charged, repelling the negatively charged PFASs. GAC's hydrophobic effects also contribute to adsorption and are independent of pH, but overall adsorption efficiency would still decline unless the solution is neutralized after the strong base treatment.[47] After adsorption, the solution could be reused for further base hydrolysis. A simple schematic of this solution can be seen in Figure 6.2.

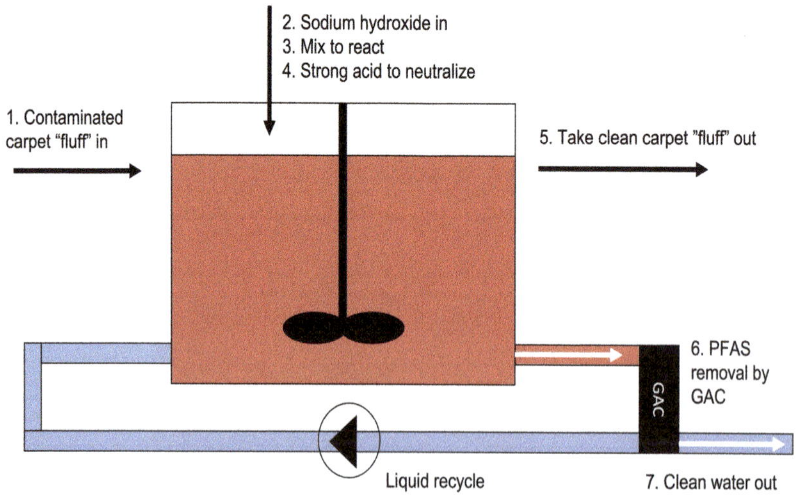

Figure 6.2 Schematic of base hydrolysis with a GAC.

6.3.1.1 Technical Performance

Base hydrolysis requires adsorption to capture released PFASs, which is not an ideal outcome (Figure 6.4). This strategy moves PFASs to a different medium, albeit in a much more concentrated and manageable form. Thermal regeneration (700 °C under a N_2 atmosphere) can potentially enable reuse of the GAC, while also mineralizing sorbed PFASs.[48] However, this regeneration process very likely requires high energy inputs and has potential to release PFASs into the air. Alternatively, saturated GAC can be stored until a better alternative is developed.

The student team chose to evaluate GAC because it is the most common type of adsorbent used for PFAS treatment, owing to its high adsorption capacity and widespread availability.[49] The efficacy of GAC has also been demonstrated at full-scale under complex matrices.[50] Anion exchange resins are also a possible choice of adsorbent because of their high removal capacity, especially for short-chain PFASs, where GAC can sometimes underperform.[51] However, these resins specifically target anionic PFASs, but not neutral or cationic PFASs. Minerals, biomaterials, and polymers have been studied for the removal of PFASs, and while there are many characteristics that make these materials desirable (*e.g.*, lower costs, higher selectivity, more carbon-neutral sourcing), these processes have not been demonstrated outside the laboratory and are not yet ready for full-scale application.[52] For these reasons, GAC would be a conservative and predictable strategy, although ensuring that the GAC is replaced before breakthrough of short-chain PFASs is critical for appropriate performance.

The cost and time to implement this strategy is relatively low. If the recycling facility already has a unit to wash the fiber, then very little capital investment would be needed. Sodium hydroxide and GAC are also relatively cheap, although more in-depth cost analyses will be needed to evaluate installation, operation, and maintenance. The simplicity of this process also makes it very feasible to implement. All the technologies are well-established and low risk, which make this strategy a cheap, fast, and reliable way to remove PFASs from carpets.

Nevertheless, base hydrolysis may damage nylon fiber quality. Empirical tests would be required to make any strong conclusions about the effect of alkaline solution on fiber quality.

6.3.1.2 Health and Environmental Criteria

There are relatively few chemicals involved in the base hydrolysis method, although using a strong base (and a strong acid to neutralize the base) introduces new hazards and potential exposures for specific worker populations (Figure 6.5). Sodium hydroxide exhibits moderate systemic toxicity and very high irritation to mucous membranes.[53,54] Hydrochloric acid exhibits low concern for carcinogenicity, mutagenicity, reproductive toxicity, developmental toxicity, neurotoxicity, and systemic toxicity, but very high concern for irritation to mucous membranes.[55,56] Potential exposures to these two compounds could be reduced through automation in a closed system.[57]

By removing PFASs from carpets during recycling in a controlled manner, the technology captures PFASs that would otherwise persist in the environment. However, the best method for handling the PFAS-saturated GAC remains an open question.

6.3.2 Base Hydrolysis with RO and Plasma

In this strategy, PFASs are again released from polymeric backbones with base hydrolysis. The PFASs in solution are then passed through an RO membrane to concentrate the PFASs. The RO permeate can be recycled back to the reactor, while the concentrate is pumped to a plasma reactor for PFAS destruction. The steps in this treatment are outlined below, and in Figure 6.3.

RO is a simple process where water is pushed through pores to exclude high molecular weight and charged compounds, generating nearly pure water in the permeate and concentrated solutions in the rejected water.[59] To push water through the membrane, pumps must generate a substantial amount of pressure to overcome the osmotic pressure generated by the solution. For example, seawater has an osmotic pressure of about 400 psi.[60] An RO facility treating seawater must generate substantially higher pressure than 400 psi to force water through the pores. Adding a strong base would greatly increase the osmotic pressure and, therefore, the cost of pumping. Furthermore, neutralization with a strong acid prior to RO would be necessary to prevent scaling or damaging the RO system. Future research should determine the minimum alkalinity needed to hydrolyze the ester bonds to reduce hazards as well as pumping costs. Although cost is a

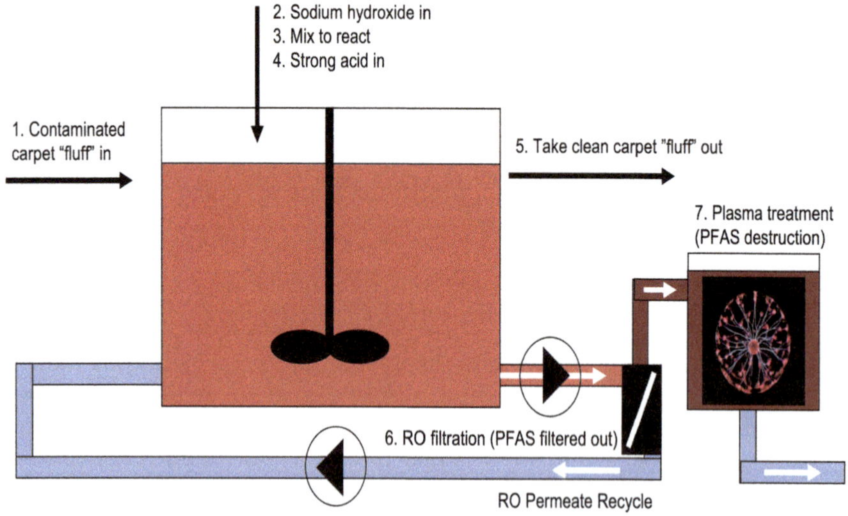

Figure 6.3 Schematic of base hydrolysis with RO and plasma.

drawback of this strategy, RO is a well-established and effective method of removing PFASs from water.[61]

Plasma is created when an extremely high electric potential is generated between two electrodes until an arc is formed.[62] Numerous plasma reactor designs have been put forward in the literature, but the most effective design for PFAS treatment uses a negative electrode in a gas headspace and a positive electrode at the bottom of the bulk liquid.[62] When the plasma is formed, it arcs from the negative electrode through an argon headspace to the liquid surface where it creates a variety of oxidizing and reducing radicals, including hydrated electrons, which are thought to be responsible for most of the reactions with PFASs. Meanwhile, a gas diffuser at the bottom of the liquid bubbles argon to concentrate the PFASs, which partition at air–water interfaces, where the short-lived hydrated electrons are most concentrated.[62] Multiple pilot scale demonstrations suggest that this is an effective and scalable technology,[63] but it is not currently commercially available.

6.3.2.1 *Technical Performance*

A substantial benefit of this strategy compared to the GAC method is that it breaks down, rather than adsorbs, PFASs (Figure 6.4). However, plasma, like most other destructive technologies, is slower in treating short-chain PFASs compared to long-chain PFASs.[62] This is likely due to the reduced hydrophobicity of shorter chains, which prevents strong air–water partitioning. This can be catalyzed with a non-fluorinated, cationic surfactant such as cetrimonium bromide (CTAB), which can bring short-chain PFASs toward the surface.[64] CTAB, a quaternary amine, is potentially toxic, but is readily

Approach	Destruction, Removal, or Nothing	Time to Implement	Reaction Speed	Energy Input	Feasibility
Base Hydrolysis + GAC	Removal	Immediately	Minutes to Hours	Low	Easy
Base Hydrolysis + RO + plasma	Destruction	Months to Years	Minutes to Hours	High	Moderate
Esterase + GAC	Removal	Months to Years	Unknown	Low	Moderate
Esterase + RO + plasma	Destruction	Months to Years	Unknown	Moderate	Moderate
Laccase mediator system	Destruction	Years	Unknown	Low	Hard

Key	Performance	High (H)	Moderate (M)	Low (L)	Unknown

Figure 6.4 Technical performance comparison of the proposed remediation strategies.

broken down in the plasma reactor.[64] Nevertheless, this method should be evaluated to prevent any inadvertent release of short-chain PFASs or quaternary amines into the environment.

This technology is relatively mature. It could be implemented relatively quickly, assuming that the necessary expertise is obtained. However, it may be slightly slower than the previous strategy because of the additional steps involved. After RO, plasma treatment takes an additional one to two hours.[64] The entire process also requires much more energy than the previous strategy. Both RO and plasma require electricity, with RO being notoriously energy intensive.[65]

6.3.2.2 *Health and Environmental Criteria*

RO and plasma are physical processes. Therefore, they require no additional chemicals. The student team included hydrofluoric acid (HF) in their evaluation since any PFAS destruction technology may yield HF. In the event that HF is produced, it may disassociate quickly into fluoride and water. Nevertheless, a conservative hazard assessment is presented in Figure 6.5.

While plasma may not introduce chemical hazards, it does have significant physical hazards. These reactors operate at a peak amperage of approximately 100 amps, which is lethal.[68] Strict controls would need to be implemented to ensure the safe operation of a plasma reactor.

Strategy	Constituent	C/M	D/R	E	S	N	I	AT	B
Base Hydrolysis + GAC	Sodium hydroxide	DG	DG	DG	M[53]	DG	VH[53,54]	DG	DG
	Hydrochloric acid	L[55]	L[58]	DG	L[58]	L[58]	VH[56]	L[58]	L[58]
Base Hydrolysis + RO + Plasma	Sodium hydroxide	DG	DG	DG	M[53]	DG	VH[53,54]	DG	DG
	Hydrochloric acid	L[55]	L[58]	DG	L[58]	L[58]	VH[56]	L[58]	L[58]
	Hydrofluoric acid	L[66]	M[67]	M[67]	H[67]	H[67]	VH[67]	M[67]	VH[67]
Esterase Replacement	Porcine liver esterase	DG	DG	DG	DG	DG	DG	DG	DG
Laccase Mediator System	Copper(II) sulfate	DG	M[75]	H[14]	DG	DG	H[75]	VH[75]	DG
	1-hydroxybenzotriazole	DG	DG	DG	DG	DG	P[77]	DG	DG
	Laccase	DG	DG	DG	DG	DG	H[76]	DG	DG
	Hydrofluoric acid	L[66]	M[67]	M[67]	H[67]	H[67]	VH[67]	M[67]	VH[67]

Key	High (H) / Very High (VH)	Moderate (M)	Low (L)	Potential concern (P)	Data gap (DG)
Carcinogenicity or mutagenicity (C/M), developmental or reproductive toxicity (D/R), endocrine toxicity (E), systemic toxicity and organ effects (S), neurotoxicity (N), eye and skin irritation (I), aquatic toxicity (AT), and bioaccumulation potential (B)					

Figure 6.5 Comparative hazard assessment of constituent chemicals of the proposed remediation strategies: base hydrolysis approach,[53–56,58] base hydrolysis + RO + plasma approach,[53–56,58,66,67] esterase replacement for base hydrolysis approach, and laccase treatment approach.[14,66,67,75–77]

6.3.3 Esterase Hydrolysis

Using a strong base and strong acid has certain drawbacks, as there are known hazards associated with their use. Strong chemicals can also damage high-value fibers or equipment. Running an RO system can be both energy-intensive and financially costly. As an alternative to base hydrolysis, the student team explored esterase enzymes, which are capable of hydrolyzing ester bonds found in (meth)acrylate linkages, thus breaking down the PFASs.

Researchers have documented esterase hydrolysis of monomeric and polymeric (meth)acrylate linkages for many decades as a result of dentistry materials research (see, for instance, Bean *et al.* 1994).[69] The porcine liver enzyme (PLE) is a particularly versatile esterase frequently cited in the literature due to its non-selective hydrolysis of ester bonds.[70] Nevertheless, it is unclear if PLE will be able to hydrolyze ester bonds if (meth)acrylate is attached to large compounds such as long-chain PFAAs. The bulky electron cloud associated with the carbon–fluorine tail of these compounds may prevent an enzyme from accessing the ester bond. While this shielding may occur, the evidence suggests that fluorine substitutions increase the rate of ester hydrolysis. If PLE can physically access the bond, the reaction will readily occur. A simple laboratory experiment could help determine if any PFASs are released from side-chain fluorinated polymers when exposed to PLE.

6.3.3.1 Technical Performance

Using an esterase instead of a strong base would increase the timeline for implementing the technology (Figure 6.4). Determining the feasibility and the optimum conditions for the esterase hydrolysis reaction to occur using PLE would take additional time and research. If the reaction with PLE proves to be too slow to be commercially viable or fails to release the fluorinated side-chains from the carpet treatment polymers, identifying an alternative enzyme would be time-consuming.

A substantial advantage of esterase replacement is its impact on energy usage in the RO and plasma membrane strategy. By eliminating the need for a strong base and acid, the osmotic pressure of the reaction solution is minimized, which reduces the pressure necessary to push the solution through the RO membrane. As a result, the use of an esterase significantly reduces the energy and financial demands of the associated RO filtration step.

6.3.3.2 Health and Environmental Criteria

No hazard data for PLE were identified across multiple databases, resulting in data gaps for all hazard endpoints in the evaluation (Figure 6.5). The lack of hazard data should not be taken as the absence of toxicity. Additional toxicity tests are needed to more accurately characterize the possible occupational risks associated with PLE in a carpet recycling setting. As for the base hydrolysis option, exposures to PLE and possible hazardous byproducts could be reduced through automation of the reaction in a closed system.

6.3.4 Mineralization with Enzymatic Laccases

Laccases are multi-copper enzymes that catalyze the reduction of molecular oxygen to water by one-electron oxidation of substrates such as substituted phenolic compounds.[71] They are widely found in plants, fungi, insects, and bacteria, and are capable of oxidizing many aromatic and non-aromatic compounds directly.[72] Laccases can also indirectly oxidize chemicals with bulky structures or high redox potentials with the aid of mediators.[73] Mediators are low-molecular weight laccase substrates, which can be easily oxidized to form free radicals. These radicals can subsequently react with target compounds that cannot directly react with the enzyme. This laccase mediator system approach has recently been shown to transform perfluorooctane sulfonate (PFOS),[74] which is resistant to most biotic degradation. This process was shown to only work in the presence of divalent cations, such as Cu^{2+}, which is suspected to catalyze conformational changes in PFOS by interacting with the negatively charged sulfonate head group.

This strategy would involve a single-step process where laccase and a chemical mediator are added directly to the face fiber fluff in suspended water. The ideal result of this reaction would be the complete mineralization of PFASs, but the technology is still in early development. In its current state, laccase oxidation of PFASs is extremely slow and inefficient. In experiments conducted by Luo *et al.* (2018), 60% of PFOS was transformed after 162 days, and most of the transformation products were non-quantifiable transformation products rather than fluoride and carbon dioxide.[74] These transformation products likely retain some of the hazard traits of PFOS. Nevertheless, with an appropriate investment in research and green chemistry to develop this strategy, the student team believes that a laccase mediator system has the potential to be the least hazardous method while effectively mineralizing PFASs into non-toxic transformation products.

6.3.4.1 *Technical Performance*

The laccase mediator system, in its current form, is able to partially oxidize PFOS and PFOA *via* a slow reaction. No studies have evaluated laccases on other PFASs, including short-chain PFAAs or fluorotelomers. The system would need to be optimized to not just transform, but also to completely mineralize these compounds. Given the scale of research and development needed, this solution is still several years away from becoming commercially viable. However, optimization and scale-up of enzymes and biocatalysts for industrial use are burgeoning in biotechnology.[72] Techniques like high-throughput screening and enzyme engineering could produce an enzyme that aligns with the objective of PFAS destruction.[72]

Assuming that such an enzyme is discovered or engineered, there are numerous technical benefits associated with this strategy (Figure 6.4). It is a single reaction, meaning a simplified reactor configuration can be used to operate the system, and it has a smaller physical footprint and lower capital costs. It would also be very energy efficient since the reaction requires no energy input. Finally, it would actually break down the PFASs, which is a significant benefit.

6.3.4.2 Health and Environmental Criteria

In its current form, the laccase mediator system performs relatively poorly from a health and environmental perspective (Figure 6.5). This is primarily due to the use of copper(II) sulfate and the mediator, 1-hydroxybenzotriazole (HBT). Copper(II) is particularly hazardous across multiple endpoints, whereas HBT, while also a serious irritant, is also very explosive, which is an endpoint not represented in the hazard table.[75] The only hazard data available for laccase was respiratory irritation, which may be of concern among exposed workers.[76]

6.4 Application

All four strategies presented have the potential to either remove or destroy PFASs during the carpet recycling process. The first and second strategies attempt to treat the carpet fiber fluff through a base hydrolysis reaction, followed by adsorption or destruction. The last two strategies use enzymes to either replace the base hydrolysis step or directly mineralize the PFASs. These strategies were presented in order of implementation time. The first two strategies can be implemented relatively quickly, while the last two would require additional time to develop and optimize (Figure 6.4).

A summary comparison of the strengths and weaknesses of these strategies is presented in Table 6.2. Evaluation of the technical performance of these strategies suggests that there are good options that may depend on a given recycling facility's needs and capacity. Any destruction-based strategy may be practical if the cost is subsidized. On the other hand, when the priority is to install a process quickly and cheaply, then base hydrolysis with GAC is likely the best solution even though it does not destroy the PFASs.

To further illustrate how the best strategy may vary based on needs and resources, it is worth considering that the reaction rates for an enzymatic process are much slower than for a chemical or physical strategy. This may make chemical or physical processes more attractive in some applications, but reaction speed is not the only important performance metric. While enzymatic processes may currently be slower, they are extremely energy-efficient, require less energy inputs, produce less brine, and reduce long-term costs of disposal.[78]

6.5 Conclusion

This case study highlights a novel application of bioinspired design to investigate technologies and processes aimed at removing PFASs from carpets.

Table 6.2 Strengths and weaknesses of each alternative strategy.

	Strengths	Weaknesses/uncertainties
Base hydrolysis + GAC	• Effective removal of side-chain fluorinated polymers • Easiest solution to implement immediately • Cheapest solution to implement right now	• Introduces new occupational exposures to strong acids/bases and PFAS-contaminated GAC • Does not mineralize the PFASs • May degrade the nylon carpet face fibers
Base hydrolysis + RO + plasma	• Requires no additional chemicals • Fast reaction times • Mineralizes PFASs	• Requires three steps to remove PFASs • RO and plasma are more expensive to operate than GAC • Introduces new occupational hazards (high voltage)
Esterase + GAC, or esterase + RO + plasma	• Enzymes require no mediator cofactors • No known hazards • Cheap and easy to implement	• Uncertain effectiveness • Slower reaction times than base hydrolysis
Laccase mediator system	• Single-step reaction • Lower hazard than esterase substitution • Potential to mineralize PFASs • Very low energy input	• Hazardous copper and HBT mediators/cofactors • Unknown, highly fluorinated transformation products • Slow reaction time • Long latency period with implementation

This is a pressing concern because there is both a need to reduce PFAS releases into the environment and to add value to postconsumer carpets. The student team incorporated technical performance and hazard criteria in their framework, aiming to develop solutions that are effective and consider the occupational hazards associated with each proposed solution. Finally, the solutions presented are adaptable enough to suit the needs of different carpet recyclers. As discussed, with the appropriate set of incentives, the solutions are feasible and can facilitate the move toward a more circular economy.

A major limitation of this case study is the lack of robust lab or field data to validate the effectiveness of these strategies. These solutions were derived from the available scientific literature and consultations with industry experts, and their effectiveness is largely inferred based on principles. In addition, given the short timeframe of the study, the student team narrowed the scope of their research and focused on evaluating solutions that could recover nylon-based face fibers. Similarly, the effect of certain carpet variables, including carpet age, use, and other chemicals used on carpets, was considered negligible in order to reduce unnecessary complexity. These factors may influence the effectiveness or hazard profiles of the solutions, presenting another opportunity for further optimization.

References

1. *Advances in carpet manufacture*, ed. K. K. Goswami, Woodhead Publishing, an imprint of Elsevier, Duxford, 2nd edn, 2017.
2. J. Valette, R. Stamm and T. Lent, *Eliminating Toxics: Lessons for the Future of Recycling*, Health Building Network, 2017.
3. DTSC (Department of Toxics Substances Control), *Product-Chemical Profile for Carpets and Rugs Containing Perfluoroalkyl or Polyfluoroalkyl Substances*, 2019.
4. Expert Team to Support Waste Implementation (ESWI), *Study on waste related issues of newly listed POPs and candidate POPs*, European Commission, 2011.
5. J. Chen, L. Tang, L. Tang, W.-Q. Chen, W.-Q. Chen, G. F. Peaslee and D. Jiang, *Environ. Sci. Technol.*, 2020, **54**, 6908–6918.
6. California Carpet Stewardship Program (CCSP), California Carpet Stewardship Program: 2021 Annual Report, https://carpetrecovery.org/wp-content/uploads/2022/09/2021_CA_AnnualReport_ADAcompliant_FINAL.pdf (accessed 10 February 2023).
7. Carpet America Recovery Effort (CARE), California Carpet Stewardship Plan 2017–2021, https://www2.calrecycle.ca.gov/Docs/Web/110031.
8. Carpet America Recovery Effort (CARE), CARE, 2019 Annual Report, https://carpetrecovery.org/wp-content/uploads/2020/06/CARE-2019-Annual-Report-6-7-20-FINAL-002.pdf (accessed 8 January 2023).
9. Carpet America Recovery Effort (CARE), 2012 CARE Annual Report, https://carpetrecovery.org/wp-content/uploads/2014/04/2012_CARE-annual-rpt.pdf (accessed 10 December 2022).
10. Carpet America Recovery Effort (CARE), Carpet Recycling 101: Post Consumer Recycling in North America, https://carpetrecovery.org/wp-content/uploads/2014/04/Carpet_Recycling_101.pdf (accessed 10 December 2022).
11. California Carpet Stewardship Program (CCSP), Highest Recyclability Update, https://carpetrecovery.org/wp-content/uploads/2021/06/Highest-Recyclability-AC-6-30-21-Final-ADA.pdf (accessed 10 December 2022).
12. N. Jeliazkova, M. Martinov, O. Tcheremenskaia, P. Rydberg, S. Avramova, N. Kochev, V. Jeliazkov, L. Iliev and M. Networks, Toxtree, https://toxtree.sourceforge.net/ (accessed 7 January 2023).
13. P. Mukerji, J. C. Rae, R. C. Buck and J. C. O'Connor, *Toxicol. Rep.*, 2015, **2**, 130–143.
14. TEDX, TEDX (The Endocrine Disruptione eXchange) List of Potential Endocrine Disruptors, https://endocrinedisruption.org/home/ (accessed 7 January 2023).

15. T. Serex, S. Anand, S. Munley, E. M. Donner, S. R. Frame, R. C. Buck and S. E. Loveless, *Toxicology*, 2014, **319**, 1–9.
16. Registration Dossier – ECHA, https://echa.europa.eu/registration-dossier/-/registered-dossier/5615 (accessed 5 November 2022).
17. Danish Environmental Protection Agency, DK-EPA – Danish Advisory List, https://eng.mst.dk/chemicals/chemicals-in-products/assessment-of-chemicals/the-advisory-list-for-self-classification-of-hazardous-substances/ (accessed 7 January 2023).
18. Environment Canada & Health Canada, Canadian Environmental Protection Act – Environmental Registrar – Domestic Substances List, https://canadachemicals.oecd.org/Default.aspx (accessed 7 January 2023).
19. J. E. Klaunig, M. Shinohara, H. Iwai, C. P. Chengelis, J. B. Kirkpatrick, Z. Wang and R. H. Bruner, *Toxicol. Pathol.*, 2015, **43**, 209–220.
20. A. L. Luz, J. K. Anderson, P. Goodrum and J. Durda, *Regul. Toxicol. Pharmacol.*, 2019, **103**, 41–55.
21. S. E. Loveless, B. Slezak, T. Serex, J. Lewis, P. Mukerji, J. C. O'Connor, E. M. Donner, S. R. Frame, S. H. Korzeniowski and R. C. Buck, *Toxicology*, 2009, **264**, 32–44.
22. H. Iwai and A. M. Hoberman, *Int. J. Toxicol.*, 2014, **33**, 219–237.
23. C. P. Chengelis, J. B. Kirkpatrick, N. R. Myers, M. Shinohara, P. L. Stetson and D. W. Sved, *Reprod. Toxicol.*, 2009, **27**, 400–406.
24. European Chemicals Agency (ECHA), Summary of Classification and Labelling for 'Undecafluorohexanoic acid' (CAS 307-24-4), https://echa.europa.eu/information-on-chemicals/cl-inventory-database/-/discli/details/125329 (accessed 7 January 2023).
25. J. M. Conder, R. A. Hoke, W. De Wolf, M. H. Russell and R. C. Buck, *Environ. Sci. Technol.*, 2008, **42**, 995–1003.
26. National Center for Biotechnology Information, PubChem Compound Summary for CID 9777, Heptafluorobutyric acid, https://pubchem.ncbi.nlm.nih.gov/compound/9777 (accessed 7 January 2023).
27. Office of Environmental Health Hazard Assessment, The Proposition 65 List, https://oehha.ca.gov/proposition-65/proposition-65-list (accessed 7 January 2023).
28. Pharos, Perfluorooctanoic acid (PFOA, C-8), https://pharosproject.net/chemicals/2012735 (accessed 7 January 2023).
29. International Labour Organization (ILO), ICSC 1613 – Perfluorooctanoic Acid, https://www.ilo.org/dyn/icsc/showcard.display?p_version=2&p_card_id=1613 (accessed 7 January 2023).
30. L. Mhadhbi, D. Rial, S. Pérez and R. Beiras, *J. Environ. Monit.*, 2012, **14**, 1375–1382.
31. G. T. Ankley, P. Cureton, R. A. Hoke, M. Houde, A. Kumar, J. Kurias, R. Lanno, C. McCarthy, J. Newsted, C. J. Salice, B. E. Sample, M. S. Sepúlveda, J. Steevens and S. Valsecchi, *Environ. Toxicol. Chem.*, 2021, **40**, 564–605.
32. ATSDR (Agency for Toxic Substances and Disease Registry), *Toxicological profile for Perfluoroalkyls*, U.S. Department of Health and Human Services, Public Health Service, Atlanta, GA, 2021.

33. Pharos, Perfluoropentanoic Acid (PFPEA, C-5), https://pharosproject.net/chemicals/2012918 (accessed 7 January 2023).
34. Safe Work Australia, Hazardous Chemical Information for nonafluoropentanoic acid, http://hcis.safeworkaustralia.gov.au/HazardousChemical/Details?chemicalID=5419 (accessed 7 January 2023).
35. European Chemicals Agency (ECHA), Substance Infocard: Perfluorononan-1-oic acid, https://echa.europa.eu/substance-information/-/substanceinfo/100.006.184 (accessed 7 January 2023).
36. J.-H. Qu, C.-C. Lu, C. Xu, G. Chen, L.-L. Qiu, J.-K. Jiang, S. Ben, Y.-B. Wang, A.-H. Gu and X.-R. Wang, *Environ. Toxicol. Pharmacol.*, 2016, **45**, 150–157.
37. W. S. Chambers, J. G. Hopkins and S. M. Richards, *Front. Toxicol.*, 2021, **3**, 732436.
38. European Chemicals Agency (ECHA), Substance Infocard: Heptadecafluorooctane-1-sulphonic acid, https://echa.europa.eu/substance-information/-/substanceinfo/100.015.618 (accessed 7 January 2023).
39. Government of Japan, GHS – Japan, https://www.nite.go.jp/en/index.html (accessed 7 January 2023).
40. N. T. Hayman, G. Rosen, M. A. Colvin, J. Conder and J. A. Arblaster, *Chemosphere*, 2021, **273**, 129699.
41. G. Toft, B. A. G. Jönsson, C. H. Lindh, A. Giwercman, M. Spano, D. Heederik, V. Lenters, R. Vermeulen, L. Rylander, H. S. Pedersen, J. K. Ludwicki, V. Zviezdai and J. P. Bonde, *Hum. Reprod.*, 2012, **27**, 2532–2540.
42. Y. Wu, K. Romanak, T. Bruton, A. Blum and M. Venier, *Chemosphere*, 2020, **251**, 126771.
43. G. F. Peaslee, J. T. Wilkinson, S. R. McGuinness, M. Tighe, N. Caterisano, S. Lee, A. Gonzales, M. Roddy, S. Mills and K. Mitchell, *Environ. Sci. Technol. Lett.*, 2020, **7**, 594–599.
44. M. M. Cooper and M. W. Klymkowsky, in *Organic Chemistry, Life, the Universe & Everything*, Michigan State University Library, 2020.
45. V. A. Nikiforov, *Chemosphere*, 2021, **276**, 130044.
46. A. R. D. Verliefde, S. G. J. Heijman, E. R. Cornelissen, G. Amy, B. Van der Bruggen and J. C. van Dijk, *Water Res.*, 2007, **41**, 3227–3240.
47. A. Ahmadi, W. Yang, S. Jones and T. Wu, *Adv. Compos. Hybrid Mater.*, 2018, **1**, 591–601.
48. N. Watanabe, M. Takata, S. Takemine and K. Yamamoto, *Environ. Sci. Pollut. Res.*, 2018, **25**, 7200–7205.
49. Y. Yao, K. Volchek, C. E. Brown, A. Robinson and T. Obal, *Water Sci. Technol.*, 2014, **70**, 1983–1991.
50. J. S. Boone, C. Vigo, T. Boone, C. Byrne, J. Ferrario, R. Benson, J. Donohue, J. E. Simmons, D. W. Kolpin, E. T. Furlong and S. T. Glassmeyer, *Sci. Total Environ*, 2019, **653**, 359–369.
51. P. McCleaf, S. Englund, A. Östlund, K. Lindegren, K. Wiberg and L. Ahrens, *Water Res.*, 2017, **120**, 77–87.
52. C. T. Vu and T. Wu, *Crit. Rev. Environ. Sci. Technol.*, 2022, **52**, 90–129.

53. Pharos, Sodium hydroxide, https://pharosproject.net/chemicals/2010314 (accessed 8 January 2023).
54. National Institute for Occupational Safety and Health, Pocket Guide to Chemical Hazards – Sodium hydroxide, https://www.cdc.gov/niosh/npg/npgd0565.html (accessed 8 January 2023).
55. International Agency for Research on Cancer (IARC), *IARC Monogr. Eval. Carcinog. Risks Hum.*, 1992, **54**, 189–211.
56. Pharos, Hydrochloric acid, https://pharosproject.net/chemicals/2008687 (accessed 8 January 2023).
57. Tech Briefs, Closed System Technology Drives the Trend Toward Safer Dispensing of High-Purity Chemicals, https://www.techbriefs.com/component/content/article/tb/supplements/mct/features/articles/9261 (accessed 10 December 2022).
58. Organisation for Economic Co-operation and Development (OECD), *Hydrogen Chloride: SIDS Initial Assessment Report*, OECD, Boston, USA, 2002.
59. J. Kucera, *Reverse Osmosis*, Scrivener Publishing, Beverly, MA, 2nd edn, 2010.
60. POWER, Osmotic power from the ocean, https://www.powermag.com/osmotic-power-from-the-ocean/ (accessed 10 December 2022).
61. C. Y. Tang, Q. S. Fu, A. P. Robertson, C. S. Criddle and J. O. Leckie, *Environ. Sci. Technol.*, 2006, **40**, 7343–7349.
62. R. K. Singh, N. Multari, C. Nau-Hix, R. H. Anderson, S. D. Richardson, T. M. Holsen and S. Mededovic Thagard, *Environ. Sci. Technol.*, 2019, **53**, 11375–11382.
63. R. K. Singh, S. Fernando, S. F. Baygi, N. Multari, S. M. Thagard and T. M. Holsen, *Environ. Sci. Technol.*, 2019, **53**, 2731–2738.
64. R. K. Singh, E. Brown, S. Mededovic Thagard and T. M. Holsen, *J. Hazard. Mater.*, 2021, **408**, 124452.
65. J. Kim, K. Park, D. R. Yang and S. Hong, *Appl. Energy*, 2019, **254**, 113652.
66. Agency for Toxic Substances and Disease Registry (ATSDR), *Toxicological profile for Fluorides, Hydrogen Fluoride, and Fluorine*, U.S. Department of Health and Human Services, Public Health Service, Atlanta, GA, 2003.
67. Pharos, Hydrofluoric acid, https://pharosproject.net/chemicals/2008685 (accessed 8 January 2023).
68. OSHA Training Institute, *Construction Focus Four: Electrocution Hazards: Instructor Guide*, OSHA, 2011.
69. T. A. Bean, W. C. Zhuang, P. Y. Tong, J. D. Eick and D. M. Yourtee, *J. Biomed. Mater. Res.*, 1994, **28**, 59–63.
70. A. J. Adler and G. B. Kistiakowsky, *J. Am. Chem. Soc.*, 1962, **84**, 695–703.
71. J. Yang, W. Li, T. B. Ng, X. Deng, J. Lin and X. Ye, *Front. Microbiol.*, 2017, **8**, 832.
72. D. M. Mate and M. Alcalde, *Microb. Biotechnol.*, 2017, **10**, 1457–1467.

73. Z.-B. Guan, Q. Luo, H.-R. Wang, Y. Chen and X.-R. Liao, *Cell. Mol. Life Sci.*, 2018, **75**, 3569–3592.
74. Q. Luo, X. Yan, J. Lu and Q. Huang, *Environ. Sci. Technol.*, 2018, **52**, 10617–10626.
75. Pharos, Cupric sulfate, https://pharosproject.net/chemicals/2007489 (accessed 8 January 2023).
76. National Center for Biotechnology Information, PubChem Compound Summary for CID 3013170, Laccase, https://pubchem.ncbi.nlm.nih.gov/compound/3013170 (accessed 10 December 2022).
77. Pharos, 1-hydroxybenzotriazole, anhydrous, https://pharosproject.net/chemicals/2013554 (accessed 8 January 2023).
78. E. L. Bell, W. Finnigan, S. P. France, A. P. Green, M. A. Hayes, L. J. Hepworth, S. L. Lovelock, H. Niikura, S. Osuna, E. Romero, K. S. Ryan, N. J. Turner and S. L. Flitsch, *Nat. Rev. Methods Primers*, 2021, **1**, 46.
79. European Chemicals Agency (ECHA), Registration Dossier for 3,3,4,4,5,5,6,6,7,7,8,8,8-tridecafluorooctan-1-ol, https://echa.europa.eu/registration-dossier/-/registered-dossier/5615/4/8 (accessed 7 January 2023).

CHAPTER 7

PFASs in Building Products: Survey of Uses and Potential Alternatives

T. A. BRUTON[a] AND H. L. RAY*[b]

[a] California Department of Toxic Substances Control, Sacramento, CA, 95812, USA; [b] International Living Future Institute, Seattle, WA, 98122, USA
*Email: Hannah.ray@living-future.org

7.1 Introduction

Uses of perfluoroalkyl and polyfluoroalkyl substances (PFASs) in consumer goods such as food packaging, outdoor clothing, and cosmetics are well known. In contrast, PFAS-containing building products – a market with an annual global economic impact of $26 billion according to the American Chemistry Council[1] – have received relatively little attention. Building materials are purchased in large quantities, and PFASs used in the manufacture of building products may have a correspondingly high impact. The production of PFASs for use in building products and the eventual disposal of these building products are likely to be significant sources of PFAS releases into the environment. There is also the potential for PFASs to migrate out of building products and lead to exposure for building occupants, workers, and the surrounding environment. Limited research has been carried out to date examining the release of PFASs from building products; however, considering the toxic legacy associated with building products such as lead-based paint and polychlorinated biphenyl (PCB)-containing sealants, a precautionary approach is warranted. Strategies aimed at limiting the use of

Green Chemistry Series No. 81
Toward a PFAS-free Future: Safer Alternatives to Forever Chemicals
Edited by Simona A. Bălan, Thomas A. Bruton and Kimberly G. Hazard

Published by the Royal Society of Chemistry, www.rsc.org

PFASs that do not address building products are missing an important piece of the puzzle.

This chapter presents a survey of the various applications of PFASs in the built environment, including roofing materials, coatings, flooring, sealants and adhesives, glass, fabrics, and more. Each section also includes a discussion of possible alternatives to PFASs for that product category, although the potential hazards of these alternatives are not assessed in detail. The chapter concludes with a description of building sector initiatives aimed at identifying and increasing the use of non-PFAS alternatives. The intent of the chapter is to highlight a lesser-known group of PFAS uses and inspire the development of non-PFAS alternatives.

7.2 Methods

We obtained information about PFAS uses in building products by searching the peer-reviewed literature, government and non-governmental reports, patents, and company websites during the winter of 2020 and spring of 2021. We documented the product type, manufacturer, trade name, and the specific PFASs used. For most product categories, we were unable to determine the market share of PFAS-containing products, so within each of the product categories discussed below, the use of PFASs may be more or less common.

For each product category where PFAS-containing products were identified, we performed similar searches to gather information on non-PFAS chemicals and materials used or proposed for use in that category. Because of the large number of products and alternatives we identified, we did not attempt to catalog the potential hazards associated with each alternative.

7.3 Uses of PFASs in the Built Environment

The unique properties of PFASs make them useful in building products for functions such as weatherproofing, corrosion prevention, friction reduction, dielectric strength, and oil and water resistance (see Table 7.1).

The following sections provide information on the use of PFASs in different building product categories. Many of the PFASs used in building products are fluoropolymers,[2] a subclass of PFASs consisting of high molecular weight materials containing fluorine atoms attached to a carbon-only polymer backbone.[3] According to the American Coatings Association, applications in the building sector are one of the key drivers of growth in fluoropolymer use in the U.S.[2] Table 7.2 lists the fluoropolymers known to be used in building products, along with their acronyms.

7.3.1 Roofing

PFASs are used in four primary types of roofing materials: metal roofing, asphalt roofing, weatherproofing membranes for flat roofs, and textile-based roofs. A building's roof is in constant contact with the elements; ultraviolet (UV)

Table 7.1 Function of PFASs in different building product categories.

Product category	Function												
	Water resistance	Oil resistance	Stain resistance	Dirt resistance/easy-clean	Heat resistance	UV resistance	Reflect solar radiation	Corrosion prevention	Weather resistance	Improve durability	Levelling agent	Binder	Wetting agent
Roofing	x		x	x	x	x	x	x	x	x			
Coatings	x	x	x	x	x	x	x	x	x	x	x	x	x
Flooring	x	x	x	x							x		x
Sealants & adhesives	x	x	x							x			x
Glass	x	x	x	x	x					x			
Fabrics	x	x	x	x									
Wires & cables	x				x			x					
Tape								x					
Artificial turf													

Product category	Function										
	Anti-blocking	Deaerator	Lubrication	Improve adhesion strength	Protect pigments	Improve finish	Chemical resistance	Flame resistance	Dielectric strength	Improve dimensional stability of wood	Extrusion processing aide
Roofing					x						
Coatings	x	x			x	x	x	x		x	
Flooring											
Sealants & adhesives				x		x					
Glass											
Fabrics											
Wires & cables								x	x		
Tape			x	x				x	x		
Artificial turf											x

Table 7.2 Fluoropolymers used in building products.

Name	Abbreviation
Polytetrafluoroethylene	PTFE
Ethylene tetrafluoroethylene	ETFE
Ethylene chlorotrifluoroethylene	ECTFE copolymer
Polychlorotrifluoroethylene	PCTFE
Polyvinylidene fluoride	PVDF
Copolymers of PVDF with acrylic resins	None
Polyvinyl fluoride	PVF
Fluorinated ethylene propylene	FEP
Perfluoroalkoxy resin	None
Copolymers of vinyl fluoride and vinyl ether	FEVE
Copolymers of FEVE with acrylic resins	None
Terpolymers of tetrafluoroethylene, hexafluoropropylene, and vinylidene fluoride	None

radiation, temperature fluctuations, and precipitation can all reduce a roof's lifespan. PFASs are used to resist weathering and prolong a roof's useful life while reflecting solar radiation away from the structure, keeping the interior cool.

7.3.1.1 *Metal Roofing*

Fluoropolymers such as polyvinylidene fluoride (PVDF) and copolymers of vinyl fluoride and vinyl ether (FEVE) are used as an exterior coating on metal sheet roofs, metal shingles, flashing, and roofing nails.[2,4–7] These fluoropolymer coatings protect metal from scratching, color loss, and corrosion.[2,5] They may also contribute to cooling the structure by reflecting solar energy into the atmosphere and reducing the amount of heat entering the building.[4,5,8] Fluoropolymer coatings applied to metal roofing components are similar to those used in other types of building materials; all of these are discussed in further detail below in the section on coatings.

Polyester and silicone-modified polyester (SMP) are alternative coating materials for metal roofing components.[9] Silicone-modified polyester coatings can meet industry standard solar reflectance criteria for cool roofs.[10]

7.3.1.2 *Asphalt Roofing*

PFASs may be used in asphalt-based roofing materials as a component of granules. Asphalt shingles and asphalt flat roofs commonly incorporate mineral-based granules to add aesthetic appeal, improve weather protection, increase UV-resistance and solar reflectivity, and to aid in ballasting.[4,11] Roofing granules consist of a core of rock or mineral covered with a pigmented ceramic coating.[4,11] Specialty granules designed for high solar reflectivity can contain PFASs such as fluorinated polyurethane silane and tridecafluoro-1,1,2,2-tetrahydroocyl-triethoxysilane.[12,13] Fluoropolymer coatings may also be applied to entire shingles.[14]

Titanium dioxide is one non-PFAS alternative used for increasing solar reflectivity.[4,13]

7.3.1.3 Weatherproofing Membranes

Fluoropolymers may be used in weatherproofing membranes for flat-type roofs in residential and commercial buildings.[15] These membranes can be made of numerous materials, including synthetic rubber, polyvinyl chloride (PVC), polyolefin, or other heavy-duty thermoplastic, and they sometimes contain a fluoropolymer layer or coating.[7,16,17] The fluoropolymers aid in moisture control and solar reflectivity, as well as conferring durability and stain resistance.[15] These membranes may also be clear or opaque for use in greenhouses.[18,19] The patent literature indicates that the following fluoropolymers may be used in weatherproofing membranes: polytetrafluoroethylene (PTFE), ethylene tetrafluoroethylene (ETFE) copolymer, ethylene chlorotrifluoroethylene (ECTFE) copolymer, polychlorotrifluoroethylene (PCTFE), PVDF, copolymers of PVDF with acrylic resins, polyvinylidene fluoride (PVF), fluorinated ethylene propylene (FEP), perfluoroalkoxy resin, FEVE, copolymers of FEVE with acrylic resins, and terpolymers of tetrafluoroethylene, hexafluoropropylene, and vinylidene fluoride.[15,16]

Heating of PTFE at high temperatures releases ultrafine particles which, when inhaled, can cause a health condition known as "polymer fume fever" and lead to severe acute lung injury.[20] Thus, installation of heat-welded roofing materials containing PTFE is a possible occupational exposure concern.

Alternative coating materials for weatherproofing membranes include silicone-, urethane-, and acrylic-based products.[15,21]

7.3.1.4 Textile-based Roofing

Fluoropolymers are used to create durable and decorative textile-based roofs. PTFE or ETFE can be applied as a coating on a rigid fiberglass base or woven into strands and then made into a textile-like material. The so-called tensile roofs made of these materials are touted for their strength, durability, and low maintenance, and are often used for retractable and deployable structures such as stadiums, and for shade fabrics.[22–24]

Non-PFAS alternatives for textile roofing include silicone-coated fiberglass, PVC-coated polyester, and high-density polyethylene.[23,24] These PFAS-free alternatives reportedly have similar lifespans, fire ratings, maintenance requirements, and UV light filtering capabilities as fluorinated tensile roofs, but at a lower cost.

7.3.1.5 Other Roofing Materials

Fluoropolymer coatings such as Dura Coat XT-10, Kynar 500, and Hylar 5000 are applied to rain gutters to repel dirt, resist staining and facilitate easier cleaning.[25,26] Alternative polyester-, silicone-, and acrylic-based coatings can provide similar weather protection without PFASs.[27,28] Gutters made from other metals such as galvanized steel have weather protective qualities and do not require coatings.[29] Non-chemical alternatives are also viable options for gutter maintenance: using leaf guards and debris strainers, and increasing the gutter slope allow water to exit the system efficiently and decrease blockages.[30]

Liquid-applied fluoropolymer coatings have been proposed for concrete roofing tiles and wooden shingles to increase reflectivity, provide cooling, and resist dirt and mildew.[14,31] Titanium dioxide-based coatings are a PFAS-free alternative for coating shingles.[14]

7.3.2 Coatings

Coatings are broadly defined as exterior treatments that serve a functional purpose. PFASs are used to improve the performance of paints, metal coatings, and wood lacquers. They reportedly protect pigments, improve ease of application, increase weather resistance, and improve the finish and durability of these products.[8] A recent report found over 100 distinct PFASs used in various paints, coatings, and finishes.[32] The same report found that coatings and paints rank as the third highest category in terms of PFAS use by manufacturers in the Nordic countries.

7.3.2.1 Paints

PFASs (both polymeric and non-polymeric) can be used as additives in epoxy-, oil-, alkyd-, and acrylic-based paints to achieve specific finishes and durability requirements.[33] Fluorosurfactant PFASs are added to paints to provide anti-blocking properties and to decrease surface tension, which allows for even flow, spread, and a glossy finish.[8,33,34] They can also be added to paints to provide non-stick, "graffiti-proof", dirt and stain resistant, oil- and water-repellent, and anti-corrosive properties.[34–37] Fluorinated polysiloxanes are added to paints as a deaerator to decrease bubbling.[6]

PFASs are also used in paints as binders.[6,33] Binders are polymeric materials that bind the ingredients in the paint together or help the paint to impregnate the substrate, minimizing bubbling and peeling. Commercially available PFAS-based binders include FEVE resins sold under the trade names ZEFFLE by Daikin Chemicals[38,39] and Lumiflon by AGC Chemicals.[40]

PFASs are used in powder coating – a dry finishing process in which pigmented polymer powders are melted onto metal, wood or other surfaces.[41,42] Fluoropolymers used in powder coating include FEVE, ECTFE, PVDF, perfluoro(alkyl vinyl ether), tetrafluoroethylene perfluoromethyl vinyl ether, ETFE, and FEP.[6,43] Fluoropolymers are added to powder coatings to improve weatherability, color and gloss retention, and resistance to chemicals, impacts, flames, and corrosion.[6,43,44]

Non-fluorinated chemicals used as binders in both liquid paint and powder coats include acrylics, alkyds, epoxies, polystyrenes, polyurethanes, phenolics, vinyl, and low density polyethylene (LDPE).[6,8] PFAS-free alternatives for improving paint viscosity and spread include low aromatic mineral spirits and aromatic petroleum distillates.[45] Non-fluorinated silicone polymers and sulfosuccinates can be used as levelling and wetting agents.[8] Alternative deaerators in paints include polyacrylates, polyethers, and polysiloxanes.[6]

7.3.2.2 Metal Coatings

Many metal coating formulations employ fluoropolymer binders to increase durability.[2,38,46] Fluoropolymer coatings are used as exterior finishes for large buildings, bridges, and industrial structures, in addition to high touch metal surfaces such as elevators and sanitary fixtures.[47] Metal entrances, doors, and door components (hinges, frames, latches, handles, locks, *etc.*) may be coated with PFASs to enhance durability and thermal protection.[48–53] The addition of fluoropolymers reportedly protects metal building products against weathering and staining and increases corrosion resistance.[40] PFAS-containing coatings are also used to increase the energy efficiency of metal roofs and exterior walls (by increasing reflectivity), to prevent snow and ice from sticking to roofs and gutters, and even to aid in the penetration of coated roofing nails.[4,6] The use of PFAS coatings on metal structures may sometimes be required by building codes, as is the case for bridges in Japan.[54] Most PFAS-containing metal coatings are applied during manufacturing, but they can also be applied to building materials after installation as liquids or sprays.[2,55]

PVDF is the original and one of the most common PFASs used in metal coatings.[46,56] Other fluoropolymers reportedly used in metal coatings include PTFE, FEP, PVF, PFDF, FEVE, and ETFE copolymers.[2] Brand name fluoropolymers used in metal coatings include Kynar PVDF from Arkema, Hylar PVDF from Solvay, Lumiflon FEVE from AGC Chemicals, and ZEFFLE VDF/tetrafluoroethylene copolymer from Daikin.[2,38]

Alternatives such as polyester and silicone-modified polyester are readily available at a lower cost point.[9,57] Galvanization and anodization are effective and cost-efficient alternatives for some applications.[57–60]

7.3.2.3 Wood Lacquers

PFASs are added to wood lacquers and sealers as wetting agents and to enhance oil and water repellency and stain resistance.[61–63] Various non-polymer PFASs may also be used in sealers to increase the dimensional stability of wood[64] and as matting agents in factory applied finishes for wooden products including cabinetry.[65]

Sulfosuccinates can reportedly replace PFASs as wetting agents in varnishes.[8] The patent literature indicates that paraffin waxes and silicones can replace PFASs in wood sealers for a similar water-repellent effect.[65] Heat and chemical treatment through acetylation can increase wood's dimensional stability and rot resistance.[66–69] Plastic-wood hybrids such as Trex come with increased durability and have high recycled content[70] (which may or may not contain PFASs or other chemicals of concern). Choosing wood species with naturally higher dimensional stability is an alternate way of limiting warping when wood is exposed to moisture.[66]

7.3.2.4 Plastic Coatings

PFASs can be applied to plastic surfaces for a variety of applications. For instance, some dry erase boards utilize fluoropolymer coatings.[71,72] Fluoropolymer coatings can also be applied on structural plastics that are used instead of glass or ceramics in items like windows, bathtubs, counters, shower stalls, and doors.[73,74]

7.3.3 Flooring

7.3.3.1 Carpets

PFASs have been used extensively as stain, soil, and water repellents in carpets. This topic is discussed in Chapter 6.

7.3.3.2 Resilient and Hard Flooring

PFAS are sometimes used to incorporate stain and soil repellency properties into resilient flooring materials. In 2003, Mannington Mills advertised that its resilient flooring products would contain Teflon.[75] More recently, resilient flooring manufacturer Congoleum promoted the use of Scotchgard Protector[76] and a patent from Armstrong Flooring, Inc. mentions the use of various fluoropolymers in its stain-resistant coating.[65] Overall, PFASs have been used less frequently in the manufacturing of resilient flooring than in carpets.

A more widespread use of PFASs related to resilient and hard flooring is in after-market floor protectors, finishes, waxes, and polishes. By the early 1990s, fluorosurfactants had reportedly been universally adopted in both household and institutional floor polish systems.[77] Fluorosurfactants are still used in these types of products as leveling and wetting agents. Examples include Novec from 3M and Masurf FS-120A from Mason Chemicals,[78] and Capstone FS-65 from Chemours.[79] Fluoropolymers like PTFE are also sold to manufacturers of floor coatings and polishes as product additives.[80] Alternatives to PFASs in floor polishes are discussed in detail in Chapter 5.

7.3.4 Sealants and Adhesives

7.3.4.1 Grout, Tile, Stone, and Concrete Sealers

Porous materials such as stones, grout, unglazed tiles, and concrete are often treated with a sealer or lacquer to create a smooth, water-resistant protective barrier. Sealers are used routinely in indoor applications including stone countertops, kitchen and bathroom tilework, and stone, tile, or concrete flooring. Sealers are also used in exterior applications including patios, staircases, foundations, and parking garages. PFASs are added to sealers to increase resistance to oil, water, stains, and (for exterior surfaces) snow, ice, and graffiti.[81–85] In the case of grout, PFAS-containing sealers are sometimes included as a component in the grout mix or paste.[86–88] PFASs used in sealers for porous materials include fluorinated acrylic and urethane polymers,[85] fluoropolymers,[88] fluorinated silicon polymers,[82] and possibly others.

Exposure to aerosolized fluoropolymer resins in grout sealers has been linked to pulmonary toxicity.[89] In 2005, dozens of cases of respiratory problems, dizziness, and disorientation were linked to one such product sold in major home improvement stores.[90] This prompted the U.S. Consumer Product Safety Commission to announce a voluntary recall of the product.[90]

Non-PFAS alternatives for sealing porous materials include silicones and epoxy, polyester, phenolic, urea, or furan resins.[91–93] For countertops, choosing dense slab surfaces that do not require sealing, such as porcelain or engineered stone, is an alternative to the use of PFAS-containing sealers.[94]

7.3.4.2 Adhesives

Adhesives are used to bond two materials together. They are used in many applications in building construction, including adhering tiles, flooring, drywall, ceiling, wood-related materials, and molded structures. Fluorosurfactant PFASs are added to adhesives to increase the strength of the bond adhering materials together. Fluorosurfactants improve adherence by enhancing the penetration of adhesives into their substrates and increasing the adhesive's wettability, allowing for greater spread and contact area between the adhesive and adhered materials.[34,92,95] Commercially available fluorosurfactant formulations marketed for use in adhesives include Capstone FS-30 and FS-31.[96] Limited evidence from a patent suggests that fluorocarbon-based synthetic rubbers, known as fluoroelastomers, may be used in adhesives as well.[97] Many adhesive formulations are made with non-fluorinated chemicals.[92]

7.3.5 Glass

Fluoropolymer PFASs are used in a variety of glass building materials such as windows, doors, and mirrors. Fluoropolymers are often applied to glass surfaces as coatings, but in some instances, the fluoropolymer can form a separate layer within the glass panel.[98]

7.3.5.1 Windows, Doors, and Other Glass and Ceramic Fixtures

Common building materials such as windows, mirrors, shower doors, bathtubs and toilets may be treated with PFASs. Fluorinated coatings are used to make glass and ceramic surfaces more durable and resistant to heat and abrasion, to limit the buildup of dust and debris, and to provide "easy to clean" and anti-smudge characteristics.[47,99–101] This makes PFAS-treated glass useful in hard-to-access locations, such as building facades and solar panels on roofs. Examples of PFASs advertised for use as glass and ceramic coatings include SURECO™ 2320[47] and 3M's Easy Clean Coatings series.[99,101] Fluoropolymer–glass composites have also been patented for use in fire-resistant applications.[98]

Researchers have developed fluorine-free "omniphobic" polyurethane and polydimethylsiloxane glass coatings that repel oil and water and are also self-healing and resistant to chemical wear.[102,103] These materials may be suitable alternatives to PFASs for glass and ceramic coatings, but it is not clear if they are commercially available.

7.3.5.2 *Lightbulbs*

Fluoropolymer coatings are applied to "shatterproof" light bulbs advertised for laboratory, medical, and consumer uses.[104,105] These coatings reportedly increase the durability of the bulbs and impart non-stick and easy-clean characteristics without changing the color, temperature, or shading of the light.[104–106] Fluoropolymers are also used in electroluminescent lamps such as safety exit signs in commercial buildings.[18,95] PFAS-coated light bulbs have been known to cause respiratory distress and death in birds including poultry due to release of volatile compounds from the coating.[107]

Alternatives to PFAS-containing shatterproof light bulbs include using LED or other light sources that do not require glass enclosures.

7.3.6 Fabrics

PFASs are added to fabrics to provide them with stain-, soil-, and water-repellent properties.[108] PFASs can be applied to individual fibers or finished fabrics during manufacturing, or as post-manufacture fabric treatments. Side-chain fluorinated polymers, a type of PFASs, have been widely used to treat furniture fabrics. Commercially available fabric treatments that contain or previously contained PFASs include brands such as Teflon, Nanosphere, Scotchgard, Capstone, Crypton, Crypton Green, Nanotex, Nanotex + Durablock, and GreenShield.[109] Fabric awnings used in commercial and residential spaces can be treated with PFASs for water- and stain-proofing. Four out of five acrylic awnings measured in a German study from 2019 contained fluorotelomer alcohols and perfluorocarboxylates at levels that indicate the use of side-chain fluorinated polymers.[110] Wall and window coverings made from fluoropolymer fabrics are also available.[111–113] For example, Dupont recommends the use of its Tedlar Wallcoverings in high traffic areas such as hospitals, day care centers, and restaurants.[112]

A variety of chemical alternatives to PFASs are available for use in fabrics. Examples include silicones, titanium dioxide or silicon dioxide nanoparticles, polyurethanes, acrylic polymers, paraffin waxes, derivatives of fatty acids including beeswax and mink oil, dendrimers, silanes, and melamine-based resins.[114,115] Each chemical alternative has a unique safety and performance profile, not evaluated here. Using alternative fabrics that have tighter weaves or inherent resistance to water and oil is another way to eliminate the need for PFAS-containing treatments. Olefinic (polypropylene) fabrics are one such example.[116] The choice of fabric may have a larger impact on stain visibility than any PFAS-based stain repellency treatment.[117]

7.3.7 Wires and Cables

Electrical wires and cables (groups of wires) are typically insulated with a non-conductive plastic sheath, often made of fluoropolymer PFAS. Fluoropolymers are used in wires and cables because of their outstanding insulating properties; resistance to flame, high temperatures, and corrosion; and ability to repel moisture.[8,118] This makes them useful in high-performance applications such as data and plenum cables, radiant heated flooring, and internal wiring of appliances including air conditioner units, light fixtures, and electrical ranges.[8,119–122] Several large chemical manufacturers sell or hold patents for fluoropolymer formulations specifically designed for coating wires.[118,123–128] PFASs are also used in electrical tapes in air ducts, which are covered in the 'Tape' section.

Alternatives to PFAS-containing wire sheaths include silicone, chlorosulfonated polyethylene, PVC, polyethylene, cross-linked polyethylene, chlorinated polyethylene, ethylene vinyl acetate, thermoplastic elastomer, neoprene, ethylene–propylene rubber, nylon, polyester, and polyurethane.[8,118,128] Many of these alternatives can function at high temperatures while also being resistant to corrosion. Fluoropolymer insulation for wiring and cabling is about 2–2.5 times more expensive compared to alternative insulation materials.[8]

7.3.8 Tape

PFAS-containing tape is used in electrical work, plumbing, sealing, and other applications. One of the most well-known uses of PFASs in building projects is PTFE plumber's tape, also known as thread-seal tape, which is used to seal pipe connections and other threaded fittings.[92,129] It works by filling the spaces between interlocking pipe threads, creating a seal that prevents leaking. PTFE plumber's tape is made of 100% PTFE film, which reduces friction and prevents rust.[130,131] Liquid/paste pipe thread sealant can also contain PTFE.[132]

PTFE tape can contain residual amounts of non-polymer PFASs such as PFOA, which has been used to manufacture PTFE and has also been detected in PTFE plumber's tape.[133,134] However, PTFE plumber's tape is normally in contact only with pipe threads and not in contact with flowing water.

Silicone-based thread-seal tapes are available but less common.[135,136] Liquid/paste pipe thread sealants without PFASs are available.[137,138] Such products can be stronger and more durable alternatives to PTFE tape and are thus preferred by plumbers for permanent seals.[139]

PFAS-containing fiberglass and film tapes are used in electrical applications in drop ceilings and other open-air spaces, for example, to wrap bundles of wires that run between separate floors of commercial buildings.[140] These tapes are composed of PTFE-impregnated plastic or a fiberglass base with an adhesive layer.[140–142] PTFE or other fluoropolymers serve as an insulator and a strengthening agent and help prevent abrasion.[143] These taped bundles can be surrounded by a layer of PVDF, which, in addition to the PTFE in the wire sheath, acts as a flame retardant and protectant from corrosion and water damage.[140] Fluorinated tape is also used inside conduits so that bundles of wires can be easily removed or inserted.[142]

PTFE tape is used in the manufacturing and installation of windows, doors, vents, skylights, and other structural openings. During manufacturing, fluorinated tape is employed to hold PVC frames together and prevent physical deformities during welding.[144,145] Some windows and skylights may still have this tape holding the frame together until installation or may be installed into the structure with the tape. Flashing tape used to seal the frame of a door, window, or other opening to the wooden frame of a building may also contain PFASs.[146,147]

It is unclear how often PFAS is used in flashing tape, but this type of tape can be made from numerous non-PFAS materials, including ethylene vinyl acetate, high density polyethylene, polyester, and polyurethane.[147] The fluorinated tapes used to protect PVC frames during welding may be unnecessary.

PFAS-containing tape can also be used in flooring applications to adhere carpets and resilient flooring to the underlayment or subfloor.[148,149] Fluoropolymers are also used in the release liners (covers that protect adhesive tape before application) and are thrown away after removal.

7.3.9 Artificial Turf

Artificial or synthetic turf is used as a water- and maintenance-saving alternative to live grass landscaping. Its popularity is growing, and as of 2015 there were 11 000 artificial turf fields in the U.S. and 13 000 in Europe.[150] Artificial turf is made up of synthetic fibers resembling blades of grass that are woven into a backing material.[151] The grass-like fibers are generally made from nylon, polypropylene, or polyethylene. Granular material, known as infill, is often used to fill in space between the blades and stabilize the turf surface.[151]

Recent reports indicate that both the backing and the blades of artificial turf can contain PFASs.[152] One study reported measurable levels of total fluorine in blades from eight samples of artificial turf, suggesting that PFAS may be used as polymer processing aids during the manufacture of blades.[153] Fluoroelastomer- or fluoropolymer-based processing aids can be added to melted plastic during extrusion to maintain smooth equipment operation and prevent defects in the finished plastic pieces.[154] Such processing aids would still be present in the final turf blades or backing. The same study detected very low levels of non-polymer PFASs in two samples of artificial turf backing, 1000 to 1 000 000 times lower than concentrations reported in fibers from new carpets.[155,156] Cost-effective alternatives to fluorinated processing aids include thermoplastic polyurethane-based elastomers.[157]

7.4 Moving Toward PFAS-free Building Products

Although the building industry is large, complex, and slow to change, it has in place a robust green building movement that has driven significant improvements in buildings' environmental footprints. In contrast to most modern buildings, which consume large amounts of energy, electricity, water, and materials, green buildings are designed to minimize their negative impacts on human health and the environment. A major success of the green building movement is its ability to demonstrate significant improvements in building energy efficiency. Certified green buildings can use between 25–60% less energy[158] than standard buildings, and can emit 50% fewer[159] greenhouse gas equivalents from water consumption and solid waste management. The innovative designs and technologies that enable these improvements have slowly been adopted by the wider building industry.

In recent years, the green building community has increasingly focused on the health and environmental impacts of the materials used in buildings, sometimes referred to as "embedded toxicity."[160] As awareness of PFASs in building products grows, the green building community has the necessary tools to encourage reduction of their use. The following sections describe the ways in which different actors in the green building ecosystem are contributing to the wider adoption of non-PFAS alternatives.

7.4.1 Large Purchasers

Corporations, hospitals, universities, real estate management companies, and other large purchasers are using their purchasing power to promote healthier building materials. When large purchasers communicate their demand for products that are made without PFASs, building product manufacturers have an economic incentive to respond.

Owners can communicate their demand for products that are free of PFASs by making a public commitment or by incorporating PFAS restrictions into their specifications during the construction process. This can take the form of a global design standard such as the one used by technology company Salesforce[161] or by Harvard University,[162] or an Environmentally Preferable Purchasing policy like the one developed by the Kaiser Permanente healthcare system.[163] In addition to restrictions on PFASs and other chemicals of concern, these documents can include a requirement for manufacturers to disclose the ingredients of building products.

The stringency of large purchasers' PFAS restrictions is uneven, and there is currently low awareness in the large purchaser community that PFASs are used for more than carpets and fabric finishes. Improving the alignment between different purchasers' requirements and expanding the scope of these requirements to different types of building materials would send a stronger signal to the market about the need for alternatives to PFASs.

7.4.2 Transparency Tools

Unlike food, cleaning products, or personal care products, manufacturers of building products are not required by law to disclose their products' ingredients. In the past, this has made it difficult to know whether PFASs or other harmful chemicals were present in a product. However, in the early 2010s, two green building nonprofit organizations created standardized tools for building product manufacturers to disclose product ingredients. The tools are known as Health Product Declarations[164] and Declare Labels,[165] and over 700 manufacturers have now used these tools to publish the ingredient lists of more than 10 000 building products. The increasing availability of data on PFAS usage and function allows building owners and other stakeholders to engage in discussions with product manufacturers regarding PFASs and their alternatives.

7.4.3 Third Party Certifications

With increasing awareness of PFAS usage in building materials, green certifications for buildings and building products are beginning to include broad restrictions on this chemical class. In 2022, the International Living Future Institute, the nonprofit group that maintains the Living Building Challenge green building certification, added about 5000 PFASs to their Red List of chemicals of concern.[166] Chemicals on this Red List are to be avoided in products used for building projects seeking Living Building Challenge certification. This Red List is a prominent document in the green building movement. It is referenced by the WELL green building rating system,[167] various owners' design standards, and is watched by manufacturers in the green building industry. The addition of all PFASs to the list was a significant step in raising awareness within the community. Also in 2021 and 2022, the green product certifications Cradle to Cradle,[168] GreenSeal,[169] and GreenScreen[170] all introduced new criteria that included broad restrictions on PFASs. The Business & Institutional Furniture Manufacturers Association (BIFMA) e3 2019 standard for furniture[171] and the EPA Safer Choice standard[172] for cleaning products also track and limit the use of PFASs as a class.

Despite the restriction on the usage of PFASs by a growing number of third-party certifications, the most widely used green building product certifications have not yet incorporated references to PFASs. Wider restrictions on PFASs in building products by popular certification programs would increase demand for PFAS alternatives.

7.4.4 Other Factors

In addition to the leadership of the green building movement, the building industry is being pushed to examine its PFAS use due to several other factors. Foremost among these is increasing regulation. For instance, several U.S. states have passed bans on the use of PFASs in textile furnishings such as upholstered furniture or rugs (California, Colorado, Maine, Maryland, Vermont, and Washington).[173,174] In addition to product-specific bans, regulations aimed at improving product circularity[175] and at controlling PFAS levels in wastewater, drinking water, and at contaminated sites[176] may also impact manufacturer decisions about PFAS use. Previous regulations on chemicals commonly used in building materials significantly shifted markets, including regulation of asbestos starting in 1973,[177] lead-based paint in 1978,[178] and formaldehyde in 2018.[179] The same may eventually be true for PFASs. For a more detailed description of the current regulatory landscape pertaining to PFASs, see Chapter 2.

Liability concerns are another factor that may influence decisions about the use of PFASs in building products.[180] Insurance providers are closely watching PFAS regulations and pollution lawsuits due to the industry's previous experience with asbestos and its expensive, unanticipated insurance claims.[181] As a start, insurance providers are beginning to introduce PFAS exclusions in their general liability insurance policies.[182]

7.5 Conclusion

Building products represent a large and overlooked sector in the usage of PFASs. Alternatives to PFASs exist for many building product applications, but questions remain regarding their efficacy and possible hazard traits. As awareness of PFAS use in the building industry grows, so will the incentive for innovators to develop safer substitutes. Adoption of non-PFAS alternatives is likely to be hastened by leadership from the green building community, as well as regulatory and liability concerns.

Acknowledgements

Much of the research summarized in this chapter was performed while the authors were employed by the Green Science Policy Institute and published in that organization's report "Building a Better World: Eliminating Unnecessary PFAS in Building Materials". The authors acknowledge the valuable contributions of Seth Rojello Fernández to this research. The authors also acknowledge Jenna Tan, Tessa Wardle, and Yuning Xu for their review.

References

1. FluoroCouncil, Fluorotechnology, https://businessdocbox.com/Green_Solutions/107328049-Fluorotechnology-critical-to-modern-life.html (accessed 16 November 2022).

2. American Coatings Association, Fluoropolymer Coatings for Architectural Applications, https://www.paint.org/coatingstech-magazine/articles/fluoropolymer-coatings-for-architectural-applications/ (accessed 17 November 2022).
3. R. Lohmann, I. T. Cousins, J. C. DeWitt, J. Glüge, G. Goldenman, D. Herzke, A. B. Lindstrom, M. F. Miller, C. A. Ng, S. Patton, M. Scheringer, X. Trier and Z. Wang, *Environ. Sci. Technol.*, 2020, **54**, 12820–12828.
4. H. Akbari, P. Berdahyl, R. Levinson, S. Wiel, B. Miller and A. Desjarlais, Cool Color Roofing Materials, https://www.coolrooftoolkit.org/wp-content/uploads/2012/04/CEC-500-2006-067.pdf (accessed 17 November 2022).
5. EDCO, Infiniti Textured Metal Shake Roofing|EDCO, https://www.edcoproducts.com/products/steel-roofing/roofing-infiniti-textured-shake.html (accessed 17 November 2022).
6. L. W. McKeen, *Fluorinated coatings and finishes handbook: the definitive user's guide and databook*, William Andrew Pub, Norwich, NY, 2nd edn, 2015.
7. National Roofing Contractors Assocation, 2004-05 Low-slope Roofing Materials Guide, https://www.nrca.net/technical/librarydetail/WuZ1K0HnHuQ= (accessed 17 November 2022).
8. OECD, *Per- and Polyfluoroalkyl Substances and Alternatives in Coatings, Paints and Varnishes (CPVs)*, Environment, Health and Safety, Environment Directorate, OECD, 2022.
9. P. Berdahl, H. Akbari, R. Levinson and W. A. Miller, *Constr. Build. Mater.*, 2008, **22**, 423–433.
10. McElroy Metal, Cool Metal Roofing and Energy Star Compliance, https://www.mcelroymetal.com/green-building-solar-metal-roofing/cool-metal-roofing (accessed 17 November 2022).
11. RoofSlope, Roofing Granules & Their Role on Flat Commercial Roofs, https://roofslope.com/roofing-granules-role-flat-commercial-roofs/ (accessed 17 November 2022).
12. 3M Innovative Properties Co., *US Pat.*, US20150113901A1, 2015.
13. Certainteed LLC, *US Pat.*, US20140248467A1, 2014.
14. R. Levinson, H. Akbari, P. Berdahl, K. Wood, W. Skilton and J. Petersheim, *Sol. Energy Mater. Sol. Cells*, 2010, **94**, 946–954.
15. Arkema Inc., *US Pat.*, US20060264563A1, 2006.
16. Firestone Building Products Co LLC, *US Pat.*, US20090181216A1, 2009.
17. GAF, Kymax Coating Product Data Sheet, https://www.gaf.com/en-us/document-library/documents/productdocuments/commercialroofingsystemsdocuments/roofcoatingssystemsdocuments/roofcoatingsdocuments/pvdfroofcoatingsdocuments/unitedcoatingskymaxcoatingdocuments/Data_Sheet__United_Coatings_Kymax_Coating.pdf (accessed 17 November 2022).
18. J. Gardiner, *Aust. J. Chem.*, 2015, **68**, 13.
19. AGC Chemicals Americas, F-CLEAN™ Greenhouse Films – AGC Chemicals, https://www.agcchem.com/products/etfe-and-greenhouse-films/f-clean-greenhouse-film/ (accessed 17 November 2022).

20. C. J. Johnston, J. N. Finkelstein, P. Mercer, N. Corson, R. Gelein and G. Oberdörster, *Toxicol. Appl. Pharmacol.*, 2000, **168**, 208–215.
21. Kynar Aquatec, Cool Roof Coatings Guide Specifications, https://kynaraquatec.arkema.com/files/live/sites/hpp_kynaraquatec/files/downloads/literature-brochures/070513-cool-roof-coatings-with-kynar-aquatec.pdf (accessed 17 November 2022).
22. Uni-Systems, The En-Fold® stadium-grade retractable awning, https://www.uni-systems.com/en-fold-retractable-awning (accessed 17 November 2022).
23. FabricArchitect, Design and Material Choices, http://fabricarchitect.com/design-and-material-choices.html (accessed 17 November 2022).
24. ShadetoOrder, Important Technical Information for Custom Made Shade Sails & Structures, https://shadetoorder.com.au/technical-information/ (accessed 17 November 2022).
25. Rain Pro Construction of Yakima, LLC, About, http://www.rainproyakima.com/about.html (accessed 17 November 2022).
26. Star Building Systems, Color Chart Selections, https://www.starbuildings.com/pdfs/colorchart/STAR-colorchart_R8_CS5.pdf (accessed 17 November 2022).
27. Duro-Last, Duro-Shield™ Roof Coatings and Materials, https://duro-last.com/products/duroshield (accessed 17 November 2022).
28. Valspar Sourcing Inc., *US Pat.*, US7381472B2, 2008.
29. Home Depot, Types of Gutters, https://www.homedepot.com/c/ab/types-of-gutters/9ba683603be9fa5395fab901cb1e4465 (accessed 17 November 2022).
30. J. M. Pilcher, *US Pat.*, US6688045B1, 2004.
31. V. Landry and P. Blanchet, *Prog. Org. Coat.*, 2012, **75**, 494–501.
32. J. Glüge, M. Scheringer, I. T. Cousins, J. C. DeWitt, G. Goldenman, D. Herzke, R. Lohmann, C. A. Ng, X. Trier and Z. Wang, *Environ. Sci.: Processes Impacts*, 2020, **22**, 2345–2373.
33. V. C. Malshe, S. Elango, S. S. Bhagwat and S. S. Maghrabi, *Prog. Org. Coat.*, 2005, **53**, 212–216.
34. E. Kissa, *Fluorinated Surfactants and Repellents*, CRC Press, 2nd edn, 2001.
35. J. Branquinho, J. A. C. Alves, S. Carvalho, J. Moniz, F. Oliveira and M. H. Gil, *J. Coat. Technol. Res.*, 2009, **6**, 483–491.
36. 3M, Stain Resistant Additive and Sealer SRC-220, https://www.palmer-holland.com/Assets/User/Documents/Product/40444/5932/SRC220StnResisAppguide_DMR.pdf (accessed 17 November 2022).
37. SEI Chemical, Non-stick Anti Graffiti Coating Graffiti Proofer Anti-Stick, http://www.seichemical.com/products/GPA-proofer_anti_stick.html (accessed 17 November 2022).
38. Daikin, Binders for Coatings, https://www.daikinchem.de/products-and-performance/binders-coatings (accessed 17 November 2022).
39. Daikin, Weatherable coating resin, https://www.daikinchemicals.com/library/pb_common/pdf/tds/Coatings_resins_and_additives/ZEFFLE_GK_Series/tds-gk-580-E_ver01_Mar_2018.pdf (accessed 14 December 2022).

40. AGC Chemicals, LUMIFLON® Functions and Features, https://www.agc-chemicals.com/jp/en/fluorine/products/detail/index.html?pCode=JP-EN-F002 (accessed 17 November 2022).
41. Powder Coating Institute, What is Powder Coating?, https://www.powdercoating.org/page/WhatIsPC (accessed 17 November 2022).
42. Powder Coating Institute, Powder Coating on Wood, https://www.powdercoating.org/page/PConWood (accessed 17 November 2022).
43. IFS Coatings, Fluoropolymer powder coatings, http://www.ifscoatings.com/content/ifs-powder-products/ifs-fluoropolymers/ (accessed 17 November 2022).
44. EI Du Pont de Nemours and Co, *US Pat.*, US7597939B2, 2009.
45. Zinsser, Technical Data PaintBooster Flow Control Additive for Oil-Based Paints, https://www.rustoleum.com/~/media/DigitalEncyclopedia/Documents/RustoleumUSA/TDS/English/CBG/Zinsser/PaintBooster_TDS_Flow_Control_Additive_Oil_Base.ashx (accessed 17 November 2022).
46. K. A. Wood, *Prog. Org. Coat.*, 2014, **77**, 2140–2146.
47. AGC Chemicals, SURECO™ AF Series Fluorinated Anti-Fouling Coating, https://www.agc-chemicals.com/jp/en/products/detail/index.html?pCode=JP-EN-G018 (accessed 17 November 2022).
48. Kawneer Company, Inc., 190/350/500 Standard Entrances, https://declare.living-future.org/products/190-350-500-standard-entrances (accessed 17 November 2022).
49. Assa Abloy, HES 1006 Series Strike, https://declare.living-future.org/products/hes-1006-series-strike (accessed 17 November 2022).
50. Norton Assa Abloy, Health Product Declaration Norton 7500 Yale 4400 Door Closer by Assa Abloy, https://www.nortondoorcontrols.com/Other/Norton/Sustainability/Health%20Product%20Declaration/HPD%20NOR%207500%20Door%20Closer.pdf (accessed 17 November 2022).
51. National Guard Products, Health Product Declaration Continuous Hinge by National Guard Products, https://hpdrepository.hpd-collaborative.org/repository/HPDs/publish_234_Continuous_Hinge.pdf (accessed 17 November 2022).
52. QMI Building Metal Products Manufacturing LLC, Health Product Declaration Steel Door by QMI Building Metal Products Manufacturing LLC, https://hpdrepository.hpd-collaborative.org/repository/HPDs/publish_646_Steel_Door.pdf (accessed 17 November 2022).
53. Assa Abloy, Health Product Declaration Sargent 10 Line Cylindrical Lock by Assa Abloy, https://hpdrepository.hpd-collaborative.org/repository/HPDs/publish_11_SARGENT_10_Line_Cylindrical_Lock.pdf (accessed 17 November 2022).
54. W. Darden, Presented in part at the Transportation Research Board 95th Annual Meeting, 2016.
55. DuPont, DuPont Snow & Ice Repellent with Teflon fluoropolymer – Aerosol Safety Data Sheet, https://www.whatsinproducts.com/files/brands_pdf/1543945754.pdf (accessed 14 December 2022).

56. J. Moses, The Evolution of Coil Coatings and Paint Systems for Metal, https://www.metalconstructionnews.com/articles/the-evolution-of-coil-coatings-and-paint-systems-for-metal (accessed 17 November 2022).
57. Akzo Nobel, CERAM-A-STAR®: The industry's best SMP coil coating system, https://coilcoatings.akzonobel.com/us/ceram-a-star (accessed 17 November 2022).
58. L. Bouchama, N. Azzouz, N. Boukmouche, J. P. Chopart, A. L. Daltin and Y. Bouznit, *Surf. Coat. Technol.*, 2013, **235**, 676–684.
59. Aluminum Anodizers Council, What are the Benefits of Anodizing Aluminum?, https://www.anodizing.org/page/anodizing-benefits (accessed 17 November 2022).
60. Home Depot, Types of Nails, https://www.homedepot.com/c/ab/types-of-nails/9ba683603be9fa5395fab909c451e98 (accessed 17 November 2022).
61. Paint & Coatings Industry Magazine, Evaluating Fluoropolymer-Based Stain-Resistant Additives and Sealers, https://www.pcimag.com/articles/101825-evaluating-fluoropolymer-based-stain-resistant-additives-and-sealers (accessed 17 November 2022).
62. Daikin Global, Coatings Resins and Additives, https://www.daikin-chemicals.com/solutions/products/coatings-resins.html (accessed 17 November 2022).
63. ECOS Paints, Zero VOC Exterior & Interior Wood Stain, https://ecospaints.net/WoodShield-Wood-Stain (accessed 17 November 2022).
64. Osmose Inc., World Intellectual Property Organization, WO2006053001A3, 2007.
65. AFI Licensing LLC, *US Pat.*, US20180327606A1, 2018.
66. R. M. Rowell and W. B. Banks, *Water repellency and dimensional stability of wood*, United States Department of Agriculture, Forest Service, 1985.
67. R. M. Rowell, R. E. Ibach, J. McSweeny and T. Nilsson, *Wood Mater. Sci. Eng.*, 2009, **4**, 14–22.
68. Chemical & Engineering News, Making Wood Last Forever with Acetylation, https://cen.acs.org/articles/90/i32/Making-Wood-Last-Forever-Acetylation.html (accessed 17 November 2022).
69. Accsys, What is Acetylation?, https://www.accsysplc.com/products/what-is-acetylation/ (accessed 17 November 2022).
70. Trex, Trex *vs.* Plastic Decking, https://www.trex.com/why-trex/composite-vs-plastic-decking/ (accessed 17 November 2022).
71. Avery Dennison Corp., *US Pat.*, US7713375B2, 2010.
72. Koroseal Interior Products, Health Product Declaration Walltalkers MagRite Dry Erase Wallcovering by Koroseal Interior Products, https://hpdrepository.hpd-collaborative.org/repository/HPDs/publish_532_Walltalkers_MagRite_Dry_Erase_Wallcovering.pdf (accessed 17 November 2022).
73. Dallas Bathtub Refinishing, Polyceramic synthetic porcelain, bathtub refinishing, finish coat, https://syntheticporcelain.weebly.com/ (accessed 14 December 2022).

74. Arkema France, World Intellectual Property Organization, WO2007064529A3, 2007.
75. FloorBiz, Flooring News: Mannington Incorporates DuPont's Teflon Into Hard Surface, http://www.floorbiz.com/BizNews/NPViewArticle.asp?ArticleID=275 (accessed 17 November 2022).
76. Congoleum, AirStep: Flexible Flooring, https://www.congoleum.com/vinyl-sheet/airstep/ (accessed 14 December 2022).
77. 3M, Presenting the solutions to your problems, https://www.ag.state.mn.us/Office/Cases/3M/docs/PTX/PTX1370.pdf (accessed 17 November 2022).
78. Advanced Concrete Technologies LLC, *US Pat.*, US10322485B2, 2019.
79. Dow, Duragreen 4373 Floor Finish Polymer, https://www.dow.com/content/dam/dcc/documents/en-us/productdatasheet/27/27-23/27-2330-01-duragreen-4373-floor-finish-polymer.pdf?iframe=true (accessed 17 November 2022).
80. Micro Powders, Inc., High Performance Wax Additives: Advanced Technology with a Quality Difference, https://www.pcimag.com/ext/resources/VirtualBrochure/MicroPowder.pdf (accessed 14 December 2022).
81. Concrete Sealers USA, PS100 Fluorinated Water, Oil & Salt Repellent WB Penetrating Sealer, https://www.concretesealersusa.com/ps100-fluorinated-water-oil-salt-repellent-wb-penetrating-sealer-1-gal/ (accessed 14 December 2022).
82. Behr Process Corp., *US Pat.*, US20140010965A1, 2014.
83. Deepshield, Gold Transparent Penetrating Sealer, https://www.deepshield.com.au/Gold-Transparent-Penetrating-Sealer (accessed 17 November 2022).
84. Norkan, Laticrete L&M Petrotex – Silane Siloxane Sealer – Concrete – 5 Gallons, https://www.norkan.com/Norkan_Store/laticrete-l-m-petrotex-silane-siloxane-sealer-concrete-5-gallons-1471 (accessed 17 November 2022).
85. 3M, 3M Stain Resistant Additives and Sealers, https://multimedia.3m.com/mws/media/750888O/3m-stain-resistant-additives-and-sealers-brochure.pdf (accessed 17 November 2022).
86. Tile Council of North America, Inc., Epoxy Grout, https://www.tcnatile.com/faqs/37-epoxy-grout.html (accessed 17 November 2022).
87. Home Depot, Tilelab Surfacegard Penetrating Sealer, https://images.homedepot-static.com/catalog/pdfImages/6c/6c6aefa7-b771-4378-b4b7-8a2ace15f311.pdf (accessed 17 November 2022).
88. Questech Corporation, World Intellectual Property Organization, WO2021163718A1, 2021.
89. G. P. Daubert, H. A. Spiller, B. I. Crouch, S. A. Seiferta, K. E. Simone and S. C. Smolinske, *J. Med. Toxicol.*, 2009, **5**, 125–129.
90. Consumer Product Safety Commission, CPSC, Tile Perfect Inc. Announce Recall of Stand'n Seal Grout Sealer Due to Respiratory Problems, https://www.cpsc.gov/Recalls/2005/cpsc-tile-perfect-inc-announce-recall-of-standn-seal-grout-sealer-due-to-respiratory (accessed 17 November 2022).

91. SealWithEase, Tile Guard Silicone Grout Sealer Review, https://sealwithease.com/tile-guard-silicone-grout-sealer-review/ (accessed 17 November 2022).
92. E. M. Petrie, *Handbook of adhesives and sealants: radiation curable adhesives; biological and naturally occurring adhesives; inorganic adhesives; nondestructive testing and industrial applications*, McGraw-Hill, New York, NY, 2nd edn, 2007.
93. Gold Eagle, TriNova® Granite Sealer, https://thegoldeaglemethod.com/trinova-granite-sealer/ (accessed 18 November 2022).
94. Healthy Building Network, Countertop Materials Hazard Spectrum, https://homefree.healthybuilding.net/products/16-countertop-materials-hazard-spectrum (accessed 18 November 2022).
95. J. Glüge, M. Scheringer, I. T. Cousins, J. C. DeWitt, G. Goldenman, D. Herzke, R. Lohmann, C. A. Ng, X. Trier and Z. Wang, *Environ. Sci.: Processes Impacts*, 2020, **22**(12), 2345–2373.
96. DuPont, DuPont Capstone Fluorosurfactants, https://cms.chempoint.com/ChemPoint/media/ChemPointSiteMedia/PDF%20Docs/K-24287-1-DuPont-Capstone-Fluorosurfactants-Brochure_1.pdf (accessed 17 November 2022).
97. Dow Chemical Co., *US Pat.*, US6005053A, 1999.
98. Solutia Inc., *US Pat.*, US20060093830A1, 2006.
99. 3M, Easy Clean Coating ECC-7000 for Ceramic and Glass Substrates, https://multimedia.3m.com/mws/media/1063806O/easy-clean-coating-ecc-7000.pdf (accessed 14 December 2022).
100. Arkema Inc., *US Pat.*, US10399892B2, 2019.
101. 3M, Easy Clean Coating ECC-3000, https://multimedia.3m.com/mws/media/1063804O/easy-clean-coating-ecc-3000.pdf (accessed 18 November 2022).
102. R. DeVito, Omniphobic Coating Suite, https://web.archive.org/web/20190715083626/https://technologies.msu.edu/omniphobic-coatings (accessed 17 November 2022).
103. A. Khan, K. Huang, M. G. Sarwar and M. Rabnawaz, *Prog. Org. Coat.*, 2020, **145**, 105703.
104. Adtech, The use of FEP in shatterproof lamp coatings – Adtech, https://adtech.co.uk/about/news/advantages-shatterproof-lamp-coatings (accessed 17 November 2022).
105. Surface Solutions Group, LLC, PTFE Coated Glass Labware – Fluoroglass, https://surfacesolutionsgroup.com/coatings/fluoroglass/ (accessed 18 November 2022).
106. Orion Industries, FluoroGlass Glass Bonding Technology, https://orioncoat.com/functional-coatings/fluoroglass-coating/ (accessed 17 November 2022).
107. M. Boucher, T. J. Ehmler and A. J. Bermudez, *Avian Dis.*, 2000, **44**, 449–453.
108. California Department of Toxic Substances Control, Product-Chemical Profile for Treatments Containing Perfluoroalkyl or Polyfluoroalkyl

Substances for Use on Converted Textiles or Leathers, https://dtsc.ca.gov/wp-content/uploads/sites/31/2019/11/Product-Chemical-Profile-for-Treatments-with-PFASs.pdf (accessed 14 December 2022).
109. C. Boyd, Upholstery applications of fluorochemicals in the office furniture industry, https://dtsc.ca.gov/wp-content/uploads/sites/31/2018/10/B1_Clinton-Boyd_PFAS-Workshop_20170131.pdf. (accessed 12 January 2021).
110. T. P. Knepper and R. M. Janousedk, *Potential SVHCs in environment and products: Measurements of the presence of potential sub-stances of very high concern in the environment and in products*, Umweltbundesamt, 2019.
111. Saint Gobain, Health Product Declaration CertainTeed Decoustics LightFrame by Saint Gobain, https://hpdrepository.hpd-collaborative.org/repository/HPDs/publish_41_CertainTeed_Decoustics_LightFrame.pdf (accessed 17 November 2022).
112. DuPont, DuPont Tedlar Wallcoverings, https://www.tedlarwallcoverings.dupont.com/content/dam/dupont/amer/us/en/microsites/wallcoverings/public/documents/Technical-DuPont-Tedlar-Wallcovering-Tech-Sheet.pdf (accessed 17 November 2022).
113. E.I. du Pont de Nemours and Company, Health Product Declaration DuPont Tedlar Wallcoverings Passport I Collection by E.I. du Pont de Nemours and Company, https://hpdrepository.hpd-collaborative.org/repository/HPDs/publish_194_DuPont_Tedlar_Wallcoverings_Passport_I_Collection.pdf (accessed 17 November 2022).
114. Amec Foster Wheeler Environment & Infrastructure GmbH (now part of Wood), The use of PFAS and fluorine-free alternatives in textiles, upholstery, carpets, leather and apparel, https://echa.europa.eu/documents/10162/13641/pfas_in_textiles_final_report_en.pdf/0a3b1c60-3427-5327-4a19-4d98ee06f041 (accessed 14 December 2022).
115. California Department of Toxic Substances Control, Potential Alternatives to PFASs in Treatments for Converted Textiles or Leathers, https://dtsc.ca.gov/wp-content/uploads/sites/31/2022/05/Public-PFAS-Treatments-Alternatives-Summary_accessible.pdf (accessed 14 December 2022).
116. J. Gibbons, What Exactly is Olefin Fabric?, https://revolutionfabrics.com/blogs/gotcha-covered/what-exactly-is-olefin (accessed 17 November 2022).
117. J. LaPier, H. Ray, C. Kwiatkowski, B. Brown, A. Blum and G. Sun, *AATCC J. Res.*, 2023, **10**, 205–213.
118. Anixter, Technical Information Handbook Wire and Cable, https://www.rmelectricsolutions.com/uploads/8/7/8/4/87841322/wire___cable_technical-handbook-en-us.pdf (accessed 17 November 2022).
119. Anixter, High-Temperature Wire, https://www.anixter.com/en_us/resources/literature/wire-wisdom/high-temperature-wire.html (accessed 14 December 2022).
120. Warmup, Electric Underfloor Heating, https://www.warmup.com/underfloor-heating/electric (accessed 17 November 2022).

121. Warmzone, FoilHeat – Low Profile Radiant Heated Floors from Warmzone, https://www.warmzone.com/floor-heating/foilheat/ (accessed 17 November 2022).
122. Winsupply, Inc., 1140 W 240 V 57.1 × 1.8 in. Advanced Fluoropolymer (ETFE) Insulation Cut and Turn Foil Heating System|Winsupply, https://www.winsupplyinc.com/warmup-heating-equipment/_/A-05369907 (accessed 17 November 2022).
123. Asahi Glass Co Ltd, *US Pat.*, US9711257B2, 2017.
124. Daikin Industries Ltd, *US Pat.*, US9012580B2, 2015.
125. Chemours, Tefzel ETFE 750 Fluoroplastic Resin, https://www.teflon.com/en/-/media/files/teflon/tefzel-750-appliance-wire-product-info.pdf (accessed 17 November 2022).
126. Dacon, Teflon® Insulated Wire (FEP, PFA, ETFE), https://dacon.com/custom-wire-manufacturing/teflon-wire-fep-pfa-etfe/ (accessed 17 November 2022).
127. Solvay, Halar® ECTFE, https://www.solvay.com/en/brands/halar-ectfe (accessed 17 November 2022).
128. KrisTech, Types of Material Used for Wire Insulation, https://www.kristechwire.com/types-of-material-used-for-wire-insulation/ (accessed 17 November 2022).
129. North American Retail Hardware Association, Study Guide: Plumbing, https://yournhpa.org/BasicTraining/Plumbing/Plumbing_Study_Guide (accessed 14 December 2022).
130. R. Cauldwell, *Taunton's Plumbing Complete: Expert Advice from Start to Finish*, Taunton Press, 2009.
131. 3M, Scotch Pipe Thread Sealant Tape 48, 547, PTFE, https://multimedia.3m.com/mws/mediawebserver?mwsId=SSSSSu9n_zu8l00xlYtUnx_vOv70k17zHvu9lxtD7xtBevSSSSSS- (accessed 17 November 2022).
132. RectorSeal T Plus 2, https://rectorseal.com/rectorseal-t-plus-2-group/ (accessed 17 November 2022).
133. T. H. Begley, K. White, P. Honigfort, M. L. Twaroski, R. Neches and R. A. Walker, *Food Addit. Contam.*, 2005, **22**, 1023–1031.
134. A. E. Rodowa, E. Christie, J. Sedlak, G. F. Peaslee, D. Bogdan, B. DiGuiseppi and J. A. Field, *Environ. Sci. Technol. Lett.*, 2020, **7**, 156–163.
135. Rust-Oleum, LeakSeal Self-Fusing Tape, https://www.rustoleum.com/~/media/DigitalEncyclopedia/Documents/RustoleumUSA/Product-Documents/IBG%20Product%20Sell%20Sheets/English%20Sell%20Sheets/CS53510514IBGSSLeakSeal%20Tape%20Final%2072314.ashx (accessed 17 November 2022).
136. PlumbingSupply.com, Plumber's Tape, https://www.plumbingsupply.com/tapes.html (accessed 17 November 2022).
137. Hercules, Megaloc Thread Sealant Product Information, https://www.oatey.com/products/hercules-megaloc-1330684393 (accessed 14 December 2022).
138. Hercules, Megaloc Thread Sealant Safety Data Sheet, https://www.oatey.com/products/hercules-megaloc-1330684393 (accessed 14 December 2022).

139. J. Kashmer, Pipe Thread Sealant *vs.* Pipe Joint Compound, https://www.pvcfittingsonline.com/resource-center/pipe-thread-sealant-vs-pipe-joint-compound/ (accessed 17 November 2022).
140. Pennwalt Corp., *US Pat.*, US4401845A, 1983.
141. Industrial Tape & Supply Company, PTFE Tape: Saint-Gobain SGK5-05 PTFE Para-Aramid Wiring Harness Tape, https://industrialtape.com/catalog/product/2386-saint-gobain-sgk5-05-ptfe-para-aramid-wiring-harness-tape.html (accessed 17 November 2022).
142. Daikin Industries Ltd, *US Pat.*, US20030082973A1, 2003.
143. Cable Ties and More, Braided Fiberglass Lacing Tape, https://www.cabletiesandmore.com/fiberglass-lacing-tape (accessed 17 November 2022).
144. Precision Coating, Fenestration, https://www.precisioncoating-ftb.com/industries/fenestration.html (accessed 17 November 2022).
145. Skyflon, PVC Door & Window Production, http://www.skyflon.com/case/PVC_Door_Window_Production-6.html (accessed 17 November 2022).
146. 3M Innovative Properties Co., *US Pat.*, US20130139953A1, 2013.
147. EI Du Pont de Nemours and Co., *US Pat.*, US8399088B2, 2013.
148. Nitto Denko Corp., *US Pat.*, US7097892B2, 2006.
149. Tandus Centiva Inc., *US Pat.*, US9988760B2, 2018.
150. A. N. Perkins, S. H. Inayat-Hussain, N. C. Deziel, C. H. Johnson, S. S. Ferguson, R. Garcia-Milian, D. C. Thompson and V. Vasiliou, *Environ. Res.*, 2019, **169**, 163–172.
151. New York State Department of Health, Crumb-Rubber Infilled Synthetic Turf Athletic Fields, https://www.health.ny.gov/environmental/outdoors/synthetic_turf/crumb-rubber_infilled/docs/fact_sheet.pdf (accessed 14 December 2022).
152. S. Lerner, Toxic PFAS Chemicals Found in Artificial Turf, https://theintercept.com/2019/10/08/pfas-chemicals-artificial-turf-soccer/ (accessed 17 November 2022).
153. PEER, Toxic Forever Chemicals Infest Artificial Turf, https://peer.org/toxic-forever-chemicals-infest-artificial-turf/ (accessed 17 November 2022).
154. Process Aids Give Pe A High-performance Punch, https://www.plasticstoday.com/process-aids-give-pe-high-performance-punch (accessed 17 November 2022).
155. X. Guo, X. Liu, K. A. Krebs and N. F. Roache, Perflourocarboxylic Acid Content in 116 Articles of Commerce, https://cfpub.epa.gov/si/si_public_record_report.cfm?Lab=NRMRL&dirEntryId=206124 (accessed 17 November 2022).
156. Changing Markets Foundation, Testing carpets for toxics: Chemicals affecting human health and hindering the circular economy. December 2018. Available at: http://www.no-burn.org/wp-content/uploads/FINAL-Testing-Carpet-for-Toxics.pdf (accessed 17 December 2018).
157. O. Kulikov, *J. Vinyl Addit. Technol.*, 2005, **11**, 127–131.

158. World Green Building Council, The benefits of green buildings, https://www.worldgbc.org/benefits-green-buildings (accessed 22 July 2022).
159. L. Mozingo and E. Arens, *Quantifying the Comprehensive Greenhouse Gas Co-Benefits of Green Buildings*, The Center for the Built Environment, University of California, Berkeley, 2014.
160. J. Blumenthal, M. L. Diamond, G. Liu and Z. Wang, *Environ. Sci. Technol.*, 2022, **56**, 9838–9841.
161. Salesforce Real Estate Design and Sustainability, Healthy Materials in the Workplace: An Introductory Guide, https://www.salesforce.com/news/wp-content/uploads/sites/3/2020/08/sustainability-healthy-materials-workplace.pdf (accessed 17 November 2022).
162. Harvard University Sustainability, Harvard University Green Building Standards, https://green.harvard.edu/sites/green.harvard.edu/files/HarvardGreenBuildingStandards2017_0.pdf (accessed 17 November 2022).
163. Kaiser Permanente, Sustainable Procurement, https://supplier.kp.org/impactspending/what-we-do/sustainable-procurement/ (accessed 17 November 2022).
164. Health Product Declaration Collaborative, Health Product Declaration Collaborative, https://www.hpd-collaborative.org/ (accessed 17 November 2022).
165. International Living Future Institute, Declare – International Living Future Institute, https://declare.living-future.org/ (accessed 17 November 2022).
166. International Living Future Institute, The Living Building Challenge (LBC) Red List 2022 Updates: A Guide for Manufacturers, https://living-future.org/wp-content/uploads/2022/05/The-LBC-Red-List-2022-and-PFAS-Guide-for-Manufacturers.pdf (accessed 13 December 2022).
167. International WELL Building Institute, WELL Standard, https://v2.wellcertified.com/en/wellv2/materials/feature/8 (accessed 17 November 2022).
168. Cradle to Cradle Products Innovation Institute, Cradle to Cradle Certified Version 4.0 Product Standard, https://cdn.c2ccertified.org/resources/STD_C2C_Certified_V4.0_FINAL_101921.pdf (accessed 17 November 2022).
169. B. Maule, Prohibiting PFAS Chemicals – Green Seal, https://greenseal.org/prohibiting-polyfluorinated-chemicals-call-for-comment/ (accessed 17 November 2022).
170. Clean Production Action, GreenScreen Certified™, https://www.greenscreenchemicals.org/certified (accessed 18 November 2022).
171. Business and Institutional Furniture Manufacturer's Association (BIFMA), e3 Furniture Sustainability Standard, https://www.bifma.org/page/e3-sustainability (accessed 13 December 2022).
172. O. US EPA, Safer Choice, https://www.epa.gov/saferchoice (accessed 13 December 2022).

173. Bryan Cave Leighton Paisner, PFAS Update, https://www.bclplaw.com/en-US/insights/pfas-update-current-state-by-state-consumer-products-regulations-july.html (accessed 13 December 2022).
174. King & Spalding, California Passes Bills Banning PFAS in Cosmetics and Textiles, https://www.jdsupra.com/legalnews/california-passes-bills-banning-pfas-in-1230995/ (accessed 13 December 2022).
175. European Commission, Circular economy action plan, https://environment.ec.europa.eu/strategy/circular-economy-action-plan_en (accessed 17 November 2022).
176. US EPA, PFAS Strategic Roadmap, https://www.epa.gov/pfas/pfas-strategic-roadmap-epas-commitments-action-2021-2024 (accessed 17 November 2022).
177. O. US EPA, EPA Actions to Protect the Public from Exposure to Asbestos, https://www.epa.gov/asbestos/epa-actions-protect-public-exposure-asbestos (accessed 17 November 2022).
178. O. US EPA, Protect Your Family from Sources of Lead, https://www.epa.gov/lead/protect-your-family-sources-lead (accessed 17 November 2022).
179. Part 770 – Formaldehyde standards for composite wood products, https://www.govinfo.gov/content/pkg/CFR-2019-title40-vol34/xml/CFR-2019-title40-vol34-part770.xml#seqnum770.8 (accessed 17 November 2022).
180. J. Gardella, PFAS Product Liability Cases – Are the Floodgates Now Open?, https://www.natlawreview.com/article/pfas-product-liability-cases-are-floodgates-now-open (accessed 17 November 2022).
181. J. Hamilton, PFAS: The Next Asbestos? – News – Tools & Intel, https://www.crcgroup.com/Tools-Intel/post/pfas-the-next-asbestos (accessed 17 November 2022).
182. R. Zarghamee and S. L. Fowler, An Update on Recent PFAS Regulation and Enforcement and the Resulting Insurance Implications, https://www.policyholderpulse.com/pfas-insurance/ (accessed 17 November 2022).

CHAPTER 8

Greener Solutions to Achieve Durable Water Repellency Without Using PFASs

E. K. COOK,*[a] M. BROWN-AUGUSTINE,[b] E. CREEL[c] AND S. RAJ[c]

[a] Department of Civil and Environmental Engineering, University of California, Berkeley, CA, 94720, USA; [b] Department of Nutritional Sciences & Toxicology, University of California, Berkeley, CA, 94720, USA; [c] Department of Chemistry, University of California, Berkeley, CA, 94720, USA
*Email: ecook21422@gmail.com

8.1 Introduction

8.1.1 Need for Durable Water-repellent Textiles

In an increasingly diverse and active world, consumers demand textiles that are water- and oil-repellent, breathable, durable, soft, and attractive. From protective gear used by firefighters and the military to warm, water-repellent gear for bikers and hikers, consumer fabrics must be impervious to everyday dirt, wear, and tear. Water repellency keeps the wearer dry and warm when exposed to rain or snow. Oil repellency increases the durability of the waterproof coating by preventing damage to the water repellency caused by sunscreen, soil, and skin oils. The combination of hydrophobic and oleophobic properties is essential for achieving durable water repellency (DWR). A large segment of the outerwear industry depends on DWR textiles. A typical DWR treatment involves applying a top coating of a multilayer "sandwich" structure consisting of fabrics and membranes. At the time of this project,

Green Chemistry Series No. 81
Toward a PFAS-free Future: Safer Alternatives to Forever Chemicals
Edited by Simona A. Bălan, Thomas A. Bruton and Kimberly G. Hazard

Published by the Royal Society of Chemistry, www.rsc.org

W. L. Gore and Associates' (Gore's) traditional membrane and fabric assembly was called GORE-TEX® laminate. This laminate consists of an inner lining, the GORE-TEX® membrane made of poly(tetrafluoroethylene) (PTFE), and an outer layer composed of the outer fabric coated with the DWR.

Gore is a major contributor to the DWR market landscape, as their DWR coatings offer some of the best omniphobic (both hydrophobic and oleophobic) properties in the textile industry. The product line GORE-TEX® is widely respected by consumers.

8.1.2 State of the Industry

As of 2017, the most effective way to achieve simultaneous water- and oil-repellency was by using fluorinated compounds to coat textiles. Fluorinated alkyl chains are unique due to the stable, hydrophobic, oleophobic, and non-reactive nature of the multiple carbon–fluorine bonds.[1] These characteristics together make fluorinated chemicals ideal components for achieving durable water repellency; indeed, no other chemicals are known to impart both hydrophobicity and oleophobicity to textiles.

Perfluoroalkyl and polyfluoroalkyl substances (PFASs) have long been used to create DWR materials; they are considered the best option for achieving both water- and oil-repellency while maintaining the durability and feel of the fabric. Although fluorinated polymers and alkyl chains are useful, certain fluorinated compounds can be described as "bad actors"–persistent, bioaccumulative, and toxic (PBT) to humans and other species.[1,2]

For the past two decades, there has been an industrial transition from using long-chain PFASs (perfluorinated alkyl chains containing eight or more carbons) to short-chain PFASs (fewer than eight carbons) in an effort to decrease the environmental persistence and toxicity of these chemicals.[3,4] Although these short-chain compounds could potentially have a shorter half-life in human bodies, there are also concerns that short-chain replacements could be more toxic.[3] Thus, there has been a growing effort to replace all PFASs and their precursors outright in DWR applications.

Greenpeace launched an international Detox campaign in 2011, urging companies in the clothing industry to eliminate all hazardous chemicals from production, focusing specifically on PFASs.[5] Together, Gore and Greenpeace defined perfluorinated and polyfluorinated chemicals of environmental concern (PFCs-EC) as chemicals that are highly fluorinated (per- or poly-fluorinated organic substances), environmentally persistent (half-life greater than two months or greater than 60 days in water or soil), and small enough to be bioavailable (capable of crossing a cell membrane or molecular weight less than 3 kDaltons). Fluorinated polymers such as polytetrafluoroethylene (PTFE) were not included in this definition. Gore partnered with Greenpeace to eliminate PFCs-EC from its products. As of 2017, their goals were to remove PFCs-EC from general outdoor products by 2020 and from specialized outdoor products by 2023.

8.1.3 W. L. Gore and Associates' Challenge

This project comprises a semester-long investigation as part of the UC Berkeley graduate course, Greener Solutions, from August to December 2017. The student project team included Marianna Augustine (molecular toxicology), Emily Cook (environmental engineering), Erin Creel (physical chemistry), Sumana Raj (physical chemistry), and John Wright (environmental health sciences). The challenge presented by W. L. Gore included goals, outcomes, and technical requirements for the UC Berkeley Greener Solutions Course project team which are outlined below.

8.1.3.1 Gore's Technical Requirements for DWR Performance

Following were Gore's specified performance metrics:

- Water resistance resulting in water contact angles equal to or greater than:
  - 115° (advancing angle)
  - 95° (receding angle)
- An oil repellency score of at least two but preferably greater than or equal to five, as outlined in the American Association of Textile Chemists and Colorists (AATCC) 118-2007
- Ability to maintain performance standards after 20 home laundry wash cycles with detergent
- Lack of negative odor
- Color, texture, and feel of final product undiminished
- Materials exhibit low inherent toxicity and are not bioaccumulative
- Be capable of high volume manufacturing

8.1.3.2 Gore's Technical Requirements for Human and Environmental Health

Gore specified that the alternative must have low inherent toxicity, be non-bioaccumulative, and must lack components that are persistent. The project team assessed bioaccumulation using criteria described by the United States Environmental Protection Agency (EPA) and the European Union's Registration, Evaluation, Authorisation and Restriction of Chemicals (REACH). REACH and the European Chemicals Agency (ECHA) prioritize substances that are considered PBT on a "substances of high concern list" as described in Section 3.2 of ANNEX XIII.[6]

The main characteristics of PBT substances are (1) high potential to bioaccumulate (high bioconcentration factor (BCF)), (2) high octanol–water partitioning coefficient (K_{ow}), (3) reproductive toxicity, and (4) slow degradation.[6]

8.1.4 Approach

8.1.4.1 Understanding PFASs Used in DWR

The project team's approach began with a literature review and study of PFASs and their current use in the textile industry. Understanding the chemical structure of fluorinated alkyl chains—and their influence on the durability, oleophobicity, and hydrophobicity of PFASs—was a useful first step in exploring other less hazardous alternatives for DWR coating. The project team found that perfluoroalkyl chains are unique; fluorine is the most electronegative element and, as a result, the carbon–fluorine bond is the strongest and shortest found in nature. Thus, PFASs have low polarizability and low surface tension, allowing them to resist wetting of nearly all fluids. These traits make it nearly impossible to replicate the performance of PFASs with other types of alkyl chains or polymers.

8.1.4.2 Bio-inspired Design

Nature has found ways to repel fluids without the use of fluorine in its years of iterative design through evolution. The project team researched plants, animals, insects, and arthropods that evolved to have water- and/or oil-repellent properties. Each team member focused on different species—lotus leaves, fish scales, snake skin, springtail exoskeletons, and rice leaves. Unexpectedly, the project team found that all these organisms, while spread across zoological and botanical families, displayed similar features. All have nano- and micro-hierarchical structuring that creates an omniphobic surface through texturing (Figure 8.1) rather than through chemical means. Additionally, many species have a type of wax coating coupled with the structuring to further improve their performance.

Hierarchical structuring allows species to decrease the contact area between the surface and the fluid by trapping air pockets between the tiny structures. The low surface contact results in minimal adhesion between the surface and fluid. In contrast, when a droplet floods the pores in a structured

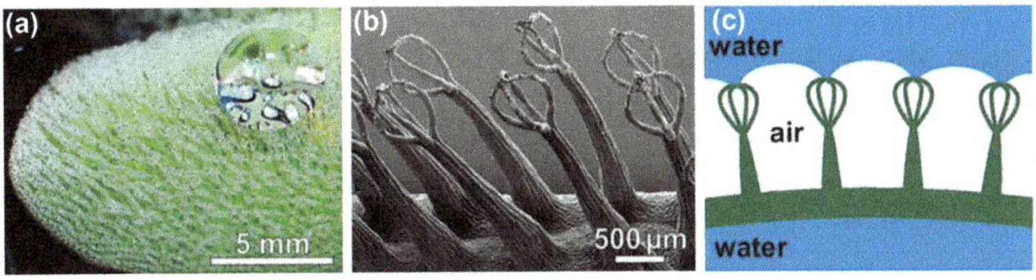

Figure 8.1 (a) Photograph of the superhydrophobic leaf of Salvinia molesta, (b) scanning electron microscopy (SEM) image of the microstructuring and nanostructuring, and (c) a schematic of how air gets trapped in this structure and repels water. Reproduced from ref. 7 with permission from Elsevier, Copyright 2015.

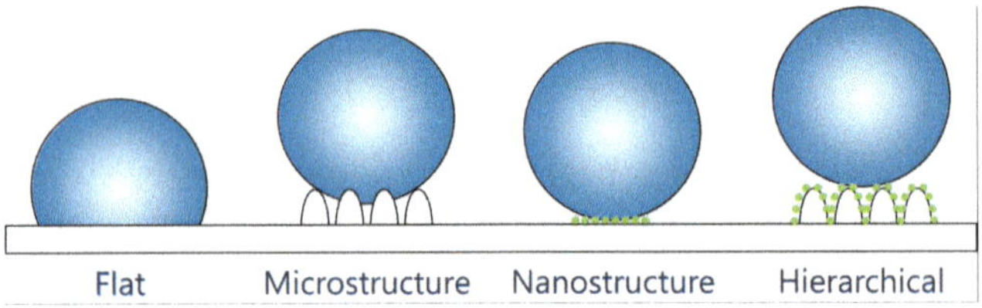

Figure 8.2 Moving from a flat to a hierarchically-structured surface decreases contact area between the fluid and the surface, increasing the effective repellency and contact angle.

surface, the increased contact area between the droplet and the surface increases adhesion (Figure 8.2).

8.1.4.3 Possibilities Ruled Out

The project team ruled out strategies that were impractical, highly hazardous, or performed poorly. For example, strictly looking at structuring to achieve an omniphobic textile surface, the project team first considered electron-beam (e-beam) lithography and 3D printing but quickly eliminated them due to their lack of durability and feasibility. Electron beams can be used to carve shapes onto the surface of a material similar to what is found in nature. These synthetic structures are effective but are easily damaged.[8] Additionally, these processes are expensive and time-consuming, making them impractical for mass production of textiles. As technology advances and becomes less expensive, these techniques may become more viable in the future.

The project team also decided to exclude all fluorinated compounds from their research. This decision was based on their preliminary hazard assessment of perfluorooctanoic acid (PFOA) and perfluorohexanoic acid (PFHxA) (see Chapter 2 for background information on PFAS hazards), the scope and goals of the Greener Solutions class, and concerns about the life cycle (production, precursor compounds, and degradation) of PFASs in the products. Although fluorinated polymers such as PTFE are not PFCs-EC (since they are not bioavailable), their production sometimes involves the use of other bioavailable PFASs such as PFOA. The project team did not investigate the potential uses of fluorinated polymers in this project to ensure that bioavailable PFASs are completely eliminated from the life cycle of the suggested alternatives.

8.1.4.4 Analyzing Potential Metrics

As the project team examined the literature, they compiled a long list of possible solutions (Section 8.2) that had shortfalls, particularly in terms of oleophobic performance. Simultaneously, the project team identified a short list of two possibilities that seemed most promising (Section 8.3). Then, the following criteria were used to narrow down the focus: the solution had to be tested on fabrics, possess high hydrophobicity, and offer a promising solution for oleophobicity.

While Gore provided specific performance standards to evaluate the alternatives, most of the accessible research reports used different, often easier-to-measure metrics. In the cases where the metric Gore requested was not available, these other metrics were used. For example, the metric Gore specified for determining a surface's repellency toward a given fluid was the static contact angle (CA), the angle at which a carefully placed liquid droplet comes in contact with a given solid surface (see Figure 8.3(a)). Fluids that readily wet the surface (-philic) have static contact angles less than 90° while fluids that are repelled (-phobic) have a static contact angle greater than 90°. A surface is super hydrophobic or oleophobic when the static contact angle is greater than 150°. In the case of water, the measured contact angle is the "water contact angle" (WCA). The WCA was the most commonly used metric in the literature due to its ease of testing in laboratory environments.

More realistic measures of the liquid repellency of a textile in a use scenario are the dynamic contact angles. Simply put, the dynamic contact angles are the contact angles between the surface of a droplet on the downhill (advancing) side of a tilted surface as well as the contact angle of the droplet on the uphill (receding) side of the surface (see Figure 8.3(b)). A highly repellent surface will have approximately equal advancing and receding angles, indicating little tailing of the fluid droplet as it is distorted by gravity.

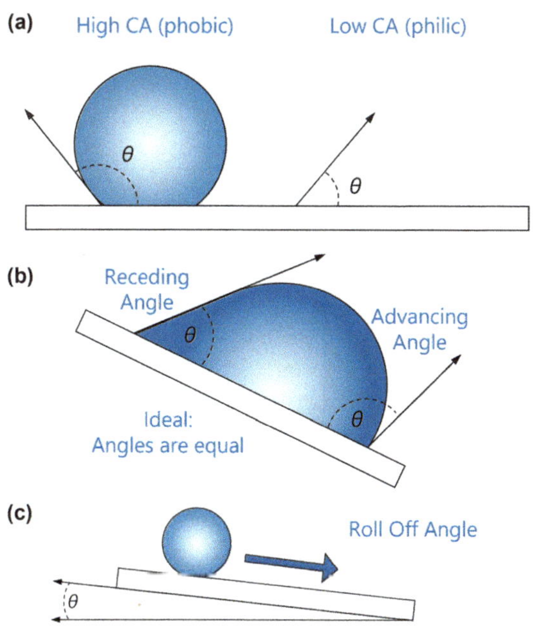

Figure 8.3 Metrics for liquid repellency are (a) static contact angle (CA), (b) dynamic contact angles, and (c) roll off angle.

A simple metric that encompasses these two angles is contact angle (CA) hysteresis (CAH). CAH is the mathematical difference between the advancing and receding angles, so a -phobic surface ideally has a 0° CAH.

Finally, the roll-off angle (see Figure 8.3(c)) is the degree to which a surface must be tilted before the liquid droplet slides off. The roll-off angle is important in textile self-cleaning applications because the droplet aids in removing dirt and oils from the surface as it is shed.

8.2 Long List of Possible Alternatives

8.2.1 Background

There are myriad ways to achieve hydrophobicity and/or oleophobicity using structure and/or surface chemistry. The following "long list" of possible solutions included strategies that needed significant improvement to achieve Gore's DWR goals. While some strategies could be used to enhance the liquid phobicity of another strategy, they may not be sufficient on their own. Many alternatives on this list managed to achieve high hydrophobicity but minimal or no oleophobicity. Although these strategies failed to meet at least one of Gore's performance goals or hazard criteria, these alternatives suggested new non-fluorinated routes to pursue.

8.2.2 Octadecylamine-grafted Nylon

A simple method for improving the hydrophobicity of nylon 6,6—one of the most common types of nylon used in textiles—is to graft aliphatic chains to the surface. One study used poly(acrylic acid) to functionalize nylon and octadecylamine as the aliphatic chain source.[9] Because nylon 6,6 has chains of six unfunctionalized carbons between its nitrogen-containing functional groups, it is relatively unreactive. Poly(acrylic acid) is highly functionalized with carboxylic acid groups, so grafting this polymer to nylon increases the binding sites for aliphatic chains. Octadecylamine binds to poly(acrylic acid) in an acid–base reaction between the amine and carboxylic acid groups. This process imparts hydrophobicity to the fabric through the 18-carbon chains of the octadecylamine coating.

However, octadecylamine's aliphatic carbon chain is highly oleophilic. Extensive modifications to improve the oleophobicity of this strategy are required to make it viable for DWR. The poly(acrylic acid) functionalization of nylon 6,6 may be useful in other strategies to increase the binding sites for other chemicals.

8.2.3 Paraffin-wrapping

Inspired by the company Schoeller and their technology ecorepel®, the project team investigated the water-repellent properties of paraffins. On their ecorepel® website, Schoeller described their sustainable technology as being "based on long paraffin chains that wrap themselves in a spiral fashion around the individual fibers".[10] The solution is abrasion resistant, breathable, dirt-repellent, and the feel is not negatively affected.

To elucidate the mechanisms of this technique, the project team identified new research that reported using various chain-length paraffins on textiles to some success.[11] Paraffins were synthesized using the monomers dodecyl acrylate, tetradecyl acrylate, hexadecyl acrylate, octadecyl acrylate, and docosyl acrylate and various emulsifiers and co-emulsifiers. This solution, autoclaved and diluted, was used to treat various fabrics with the dip and nip method.[11] The resulting paraffinic coating lowered the surface tension on PET, nylon, and cotton and achieved highly hydrophobic properties.[11] The study only investigated water contact angles and surface tension; no results were provided related to durability, abrasion resistance, washability, or oleophobicity.

8.2.4 Multifilament Fibers

Yarn used to manufacture fabric can have a variety of structures, such as monofilament, multifilament, and calendered. In a monofilament textile, the yarn is made of a single thread in contrast to a multifilament textile where the yarn is composed of a bundle of threads woven together as if they were one (Figure 8.4(a)). The structure of the yarn can also be modified after the fabric is woven. For example, calendered fabric has undergone a process where it is passed through a series of rollers to squeeze and flatten the fabric to make it more lustrous (Figure 8.4(b)).[9]

One study compared the hydrophobicity of octadecylamine- or PFOA-modified nylon with various fabric weave structures. The multifilament textiles had much higher hydrophobicity than the monofilament or calendered monofilament textiles (Figure 8.4(c)) due to the microstructuring of the fabric surface.[9]

Microstructuring through the weave of the fabric only improves existing hydrophobicity or oleophobicity; the textile must have some degree of inherent phobicity in order to reap the benefits. This is one way to improve hydrophobic fabrics such as polyester, polyolefin, and vinyl.

8.2.5 Microfibers

The multifilament yarn strategy described above can be improved by using smaller fibers and a smaller, tighter weave. Traditional polyester (fiber diameter greater than 15 μm) is less hydrophobic than microfiber polyester (fiber diameter less than 5 μm) (Figure 8.5) with the same surface treatment.[12] The smaller structures on the microfiber textiles mean that there are more air gaps and less contact area between a liquid and the fabric for a given drop size. Again, this improvement in fluid-phobicity is minimal and only of use when the base structure already has some fluid-phobicity.

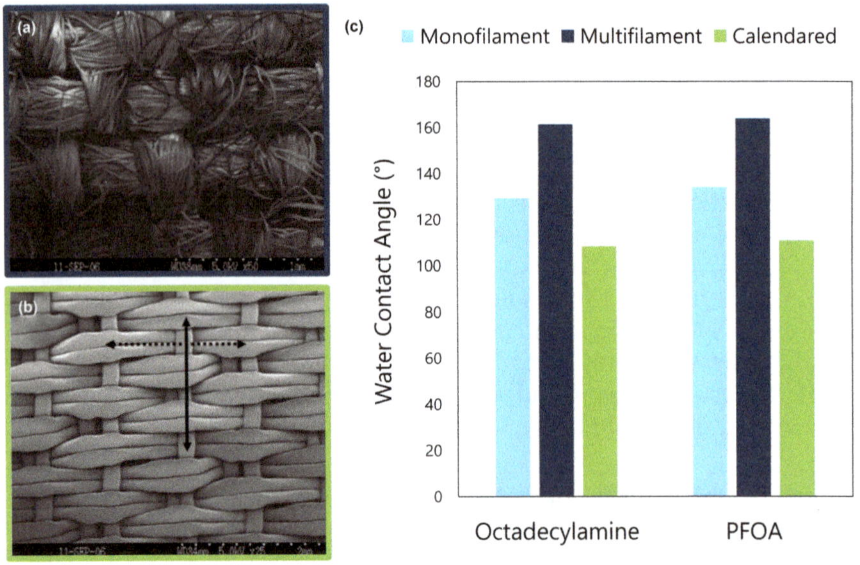

Figure 8.4 Multifilament fiber structures and water contact angles. (a) SEM of multifilament nylon, (b) scanning electron micrograph of calendered monofilament nylon, and (c) comparison in water repellency between octadecylamine- and PFOA-grafted nylon of different fiber structures, Adapted from ref. 9 with permission from American Chemical Society, Copyright 2007.

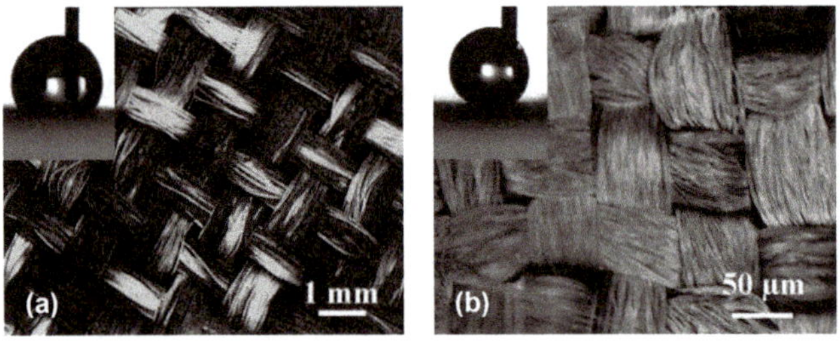

Figure 8.5 SEM of (a) traditional and (b) microfiber polyester with insets showing the static water contact angle. Adapted from ref. 12 with permission from American Chemical Society, Copyright 2006.

8.2.6 Silicone Nanofilaments

One of the most hydrophobic alternatives that the project team found in the literature was growing silicone nanofilaments on textiles. Tiny polymethylsilsesquioxane (PMSQ) hairs were grown on the surface of poly(ethylene terephthalate) (PET) fabric (Figure 8.6) in a room-temperature hydrolysis reaction, providing both physical and chemical hydrophobicity. The roll off angle

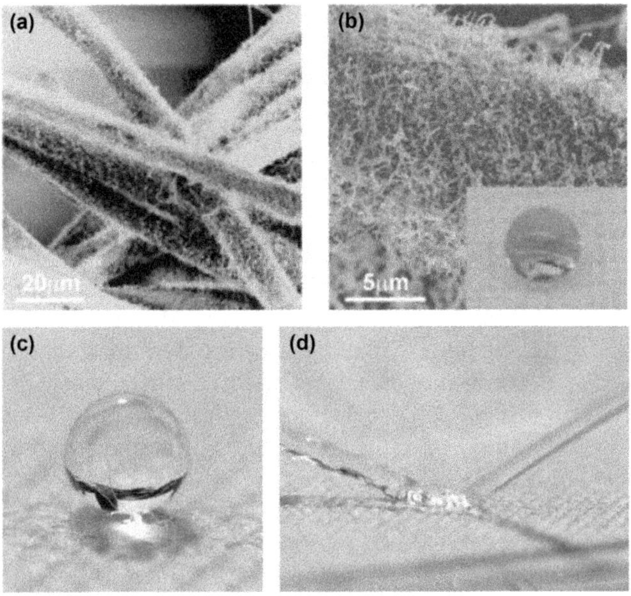

Figure 8.6 Silicone nanofilaments on PET. (a and b) SEMs of silicone nanofilaments grown on PET with insets showing a drop of water deposited on the fabric. (c) A 5 μL water droplet beads up on the surface, creating a static contact angle above the superhydrophobic threshold. (d) A jet of water bounces off the treated fabric, showing no wetting behavior. Reproduced from ref. 13 with permission from John Wiley & Sons, Copyright © 2008 Wiley-VCH Verlag GmbH & Co. KGaA, Weinheim.

of a drop of water released from a height onto the PMSQ-treated fabric was 2°.[13] The high static contact angle and a stream of water bouncing off the surface are shown in Figure 8.6.

One disadvantage of PMSQ treatment is the possibility of a very small color change in the fabric, but this change is probably imperceptible to the human eye.[13] The PMSQ nanofilaments do not affect the tensile strength or feel of the fabric, and nanofilament PET maintained high hydrophobicity after 1450 abrasion cycles despite the destruction of the nanofilaments on the surface of the fabric. The three-dimensional arrangement of fibers meant that only the outermost fibers got destroyed while the hydrophobicity was maintained by the deeper nanofilaments that remained intact. Unfortunately, washing the PMSQ-treated PET with an alkaline detergent at 30 °C resulted in water droplets wetting the fabric rather than remaining beaded on the surface. It is likely that the silicone was damaged by the alkaline detergent but could resist acidic detergents.

Due to the high oleophilicity of silicone, the silicone nanofilaments would require modification to impart oleophobicity before being used in DWR. The oleophobicity could be improved by using a dimethyldimethoxysilane coating, suggested for the silica nanosol strategy described in Section 8.3.

8.2.7 Slippery Liquid Infused Porous Structures (SLIPSs)

Another strategy for achieving omniphobicity is SLIPSs, inspired by the *Nepenthes* pitcher plant. In this alternative, instead of trapped air providing the omniphobicity, an ultrasmooth, slippery liquid lubricant was reported to provide a surface that was nonwetting to a variety of liquids and solids.[14] The liquid-like nature of the SLIPS surface provided an excellent dynamic dewetting behavior similar to the behavior of colloids. SLIPS also exhibited self-healing behaviors because the lubricant redistributed in the damaged areas.[14]

The lubricant in SLIPS was held in place by a porous solid, usually a polymer-based network for textile applications. Although the lubricant for SLIPS is often fluorinated,[14,15] some studies demonstrated fluorine-free omniphobic SLIPS using silanes. One study on a silicon wafer demonstrated a CAH of <5° and CAs of <20° for water and three organic liquids.[16] Another created an omniphobic PET fabric using poly(dimethylsiloxane) (PDMS) as the porous substrate and silicone oil as the lubricant (Figure 8.7).[17]

Although the omniphobicity of the fluorine-free SLIPS on PET is impressive, this strategy needed significant development to ensure that the lubricant stays intact during wear, abrasion, and washing.

8.2.8 Titania Nanosols

Treating a textile with titania (TiO_2) nanoparticles imparts ultraviolet (UV)-resistance, crease-resistance, self-cleaning, and antibacterial properties.

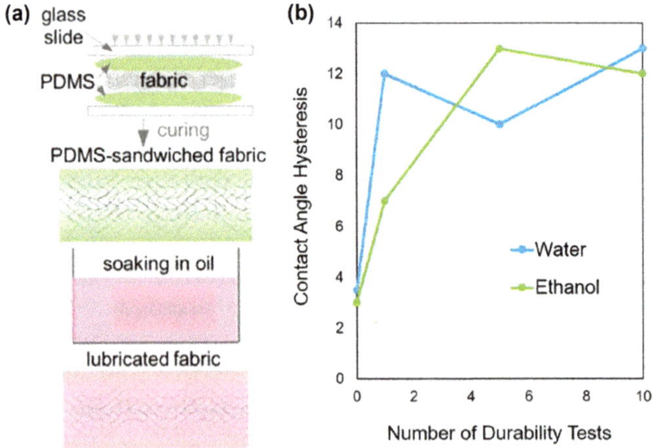

Figure 8.7 Non-fluorinated SLIPS on PET. (a) The SLIPS is created by curing a sandwich of PDMS and the fabric between two glass slides. The lubricant is added by soaking the PDMS-sandwiched fabric in silicone oil. (b) The lubricated fabric maintains low CAH for both water (blue) and ethanol (green) after up to ten durability tests. Adapted from ref. 17 with permission from American Chemical Society, Copyright 2015.

Of particular interest to DWR is the self-cleaning resulting from the photocatalytic behavior of semiconductor titania. Titania nanoparticles prepared by the sol–gel method and applied to cotton fabric using various cross-linkers achieved the top score in the AATCC Soil Release Test after they were irradiated with UV light.[18] The authors suggested that this self-cleaning effect resulted from the creation of reactive oxygen.[18] Unlike the silica nanosol strategy presented in Section 8.3, the effect of titania was photochemical rather than structural.

Stain release technologies do not necessarily meet the performance metrics for DWR as outlined by Gore. Gore wanted to avoid penetration of oily stains altogether, while stain release technologies instead ensure that the oils are removed during washing; titania nanosols could instead be combined with another strategy to enhance oleophobicity.

8.2.9 Low-hazard Bioplastic Fibers

Biodegradable fibers are a greener alternative to petroleum-based, non-biodegradable textiles such as nylon. Blends of these fibers can have a range of hydrophobicities, elasticities, and other properties.[19] For example, poly(3-hydroxybutyrate) (P(3HB)) is a polymer biosynthesized by bacteria from waste, sugars, or fatty acids.[20,21] P(3HB) can be made with varying tensile strength, brittleness, and degrees of melt stretching, and can be used to produce nanostructures.[22] While these fibers may not necessarily exhibit higher performance than conventional fibers, they do generally pose less hazards due to their low environmental persistence.

8.3 Short List of Recommendations

8.3.1 Silica Nanosol Coatings

8.3.1.1 Background

Silica (SiO_2) nanoparticles (NPs) can be applied to textiles to impart liquid repellency by adding nanostructuring to the microstructures formed using threads in fabrics. The particles are typically applied through a solution-based method. The silica sol (solution) is composed of three parts: (1) silica NPs or their precursor, (2) organosilanes, and (3) water or low molecular weight alcohols, each serving a specific function. Silica NPs are chemically inert and serve as the structural element that enhances the fluid-repellency of the fabric. The silica sol can contain either commercially available NPs or chemical precursors to silica NPs such as tetraethoxysilane (TEOS). The organosilane component of the sol serves as the water/oil repellent coating on the nanoparticles and the binder that attaches the particles to the fabric. The organosilanes consist of three short chain alkoxy groups and a non-hydrolyzable functional group bound to a central silicon atom (see Figure 8.8).

Water or low molecular weight alcohols cleave the alkoxy chains from the silicon atom through hydrolysis. The consequently available bond sites allow the silane to bind to the NPs or the fabric. The remaining functional group on the surface-bound silane decreases the surface energy of the resulting structure. This change in surface energy gives the surface its fluid-phobic properties—hydrophobicity and/or oleophobicity. The NPs only enhance the liquid repellency, making the appropriate selection of the functional group on the organosilane critical to the performance of the nanosols.

The silica sol can be applied to fabrics through spraying[23] or padding (dip and nip),[24,26] although the padding method was emphasized in this report because it is commonly used in the textile industry (including in Gore's DWR application), offers a more uniform application of the coating to the fabric, and reduces the potential to volatilize the components of the sol. In padding, the fabric on rollers is first immersed in the "dip" solution—here the nanosol—and then the excess solution is squeezed out in the "nip".[27] An optional curing step such as using a drying oven completes the coating process.

8.3.1.2 *Performance*

8.3.1.2.1 Hydrophobicity. Several studies reported high hydrophobicity using silica NPs. One such study investigating the incorporation of silica nanosols into polyester reported highly hydrophobic WCAs for static, advancing, and receding angles (Figure 8.9). These silica NPs were modified with surface alkyl groups through the presence of 3-aminopropyltriethoxysilane (APTES) and hexadecyltrimethoxysilane (HDTMS).[23] From the reaction between hydrolyzed APTES and Si–OH groups, the silica NPs aggregated to form clusters, creating a hierarchical surface structure.[23]

The resulting WCAs from the combined method of using silica NPs with the APTES and HDTMS coating exceeded Gore's goals of having advancing

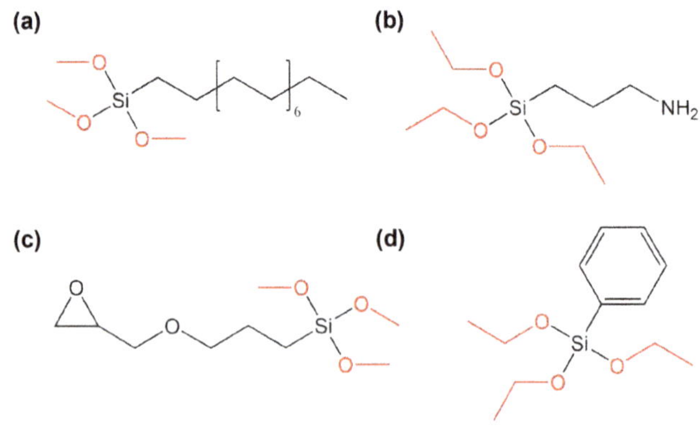

Figure 8.8 Organosilanes used in silica sols with short chain alkoxy groups indicated in red. (a) Hexadecyltrimethoxysilane (HDTMS), (b) 3-aminopropyltriethoxysilane (APTES), (c) 3-glycidoxypropyltrimethoxysilane (GPTMS), and (d) phenyltriethoxysilane (PhTES).[23–25]

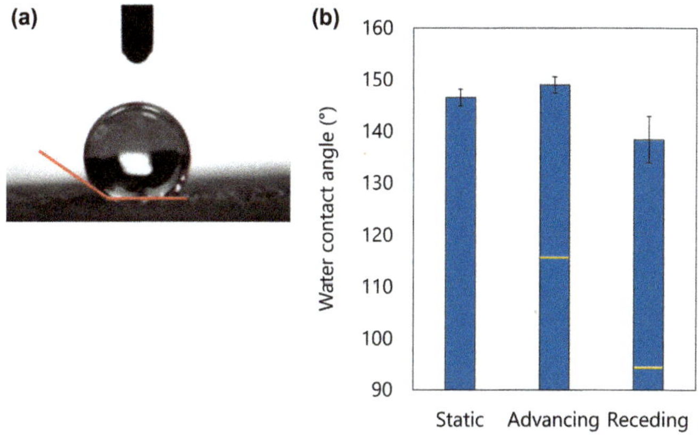

Figure 8.9 Hydrophobicity of silica nanosol coated polyester. (a) SEM image showing a high static contact angle of a water droplet indicating high hydrophobicity. (b) The silica nanosol coating imparts high static and dynamic contact angles. The yellow lines represent Gore's goals for advancing and receding angles. Image in (a) reproduced from ref. 23 with permission from Elsevier, Copyright 2016. Data in (b) from ref. 23.

and receding contact angles greater than 115 and 95°, respectively (Figure 8.9).[23] The static WCA was not quite superhydrophobic (greater than or equal to 150°) but was considered highly hydrophobic.

8.3.1.2.2 Oleophobicity. Unfortunately, nanosol fabric coatings were not very oleophobic. One study[28] looked at a number of different organosilane additives to the nanosols but found minimal improvement in the oleophobicity of the fabric (see Figure 8.10).[28] It was only with the addition of fluorinated compounds that the oleophobicity improved significantly.

Another study showed that polymerization of dimethyldimethoxysilane (DMDMS) on a surface could result in omniphobicity.[29] This layer acted like a fluid, so the interactions determining contact angles were more similar to liquid/liquid interactions compared to liquid/solid interactions. While the static contact angle of oils on this surface was not high, there was little variation between the advancing and receding contact angles, which meant that the dynamic wicking of the material was good.[29] DMDMS nanosols had not, however, been tested on fabrics.

8.3.1.2.3 Durability, Abrasion Resistance, and Washability. To test washability, authors Zhao *et al.* hand-washed their polyester fabric coated with silica nanosol in a 1% detergent solution, rinsed it with water, and air-dried it at ambient temperature.[23] This wash–dry cycle was repeated five times and the WCAs were measured again. After washing, the contact angles only decreased by a couple of degrees, resulting in static and dynamic contact angles that are still well above Gore's goals (see Figure 8.11). However, Gore had a more rigorous wash–dry standard (20 wash/dry cycles), so further tests and

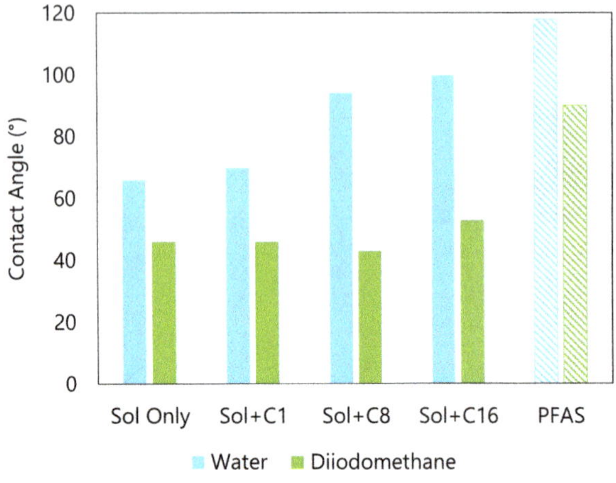

Figure 8.10 The water and organic contact angles for various organosilane additives to nanosols. Data are shown for the silica sol alone and with 1, 8, and 16 carbon chain additives. The contact angles for a nanosol-free fluorinated coating are presented for comparison. Reproduced from ref. 28 with permission from Springer Nature, Copyright 2003.

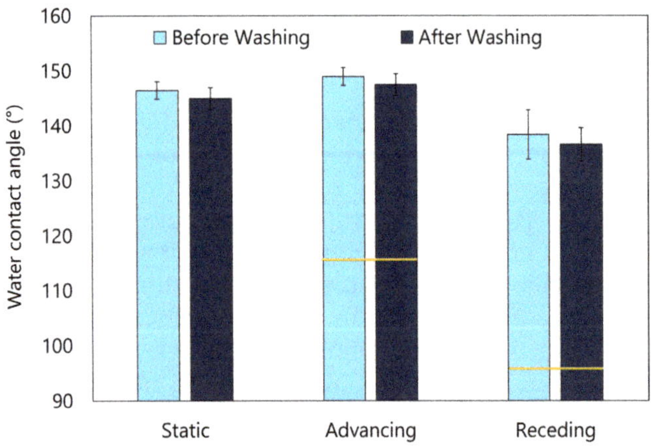

Figure 8.11 The effect of washing on WCAs for polyester coated with silica nanosol. The light blue color shows the initial WCA values; dark blue bars show the WCAs after five hand wash/dry cycles with detergent. Yellow lines represent Gore's goals for advancing and receding angle. Data from ref. 23.

studies would be needed to fully explore the washability of the silica nanosol method.

In the same study, the coated fabric was tested for abrasion resistance by soaking in a 1% detergent solution for 10 minutes, and then brushing with an Elcometer abrasion tester at a rate of 30 cycles per minute for 600 cycles.[23] After the brushing process, the fabric was rinsed with water and air-dried at ambient temperature. Only static contact angles were measured following a certain number of brushes; after starting at 145°, the static angle decreased to about 135° after 600 brushing cycles (see Figure 8.12).[23] Despite these decreases in WCAs, the final values were still greater than Gore's goals.

A different study[25] experimented with adding another silane component, 3-glycidoxypropyltrimethoxysilane (GPTMS), to improve the washability of the silica sol–gel coating. The improved washing durability for polyester was attributed to the additional epoxide group that GPTMS provided. Fabrics without GPTMS lost hydrophobicity after 50 laundering cycles; polyester with GPTMS, however, retained superhydrophobicity after 50 washing cycles.[25] Additionally, using GPTMS increased the silica particle size from about 250 nm to about 550 nm and resulted in more regularly shaped particles, possibly decreasing the bioavailability of the nanoparticles.[25]

8.3.1.2.4 Feel and Appearance. Most silica nanosol films were noted to be essentially colorless without strong odors. In fact, metal nanoparticles and other chemicals could be added into the nanosol coating to provide specific colors. Nanosol coatings appeared to have a minimal effect on breathability of the material, although they could lead to an increase in the stiffness.[30]

8.3.1.3 Hazards

8.3.1.3.1 Nanoparticle Revolution: Applications and Uncertainty. Nanoscale materials (1–100 nm) were the driving force behind the nanotechnology revolution of a new generation of materials, valued at more than a trillion dollars in the U.S. market.[31] A nanoscale substance is 10^4 times smaller than a single sheet of paper and the size of only a few blood

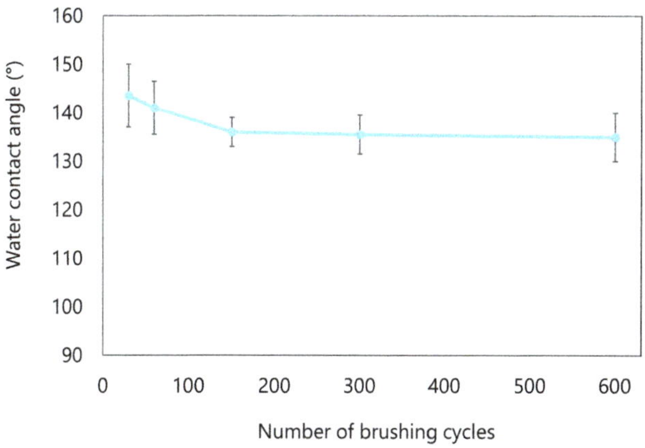

Figure 8.12 WCAs measured for polyester coated with silica nanosol during abrasion testing. The WCA decreases by only a few degrees after a defined number of brushing cycles. Data from ref. 23.

cells.[35] In the past two decades, innovations using nanotechnology in medicine, computers, water remediation, and almost all industrial sectors, have resulted in a wide range of nanoparticles entering the environment. Silicate, titanate, and zirconate nanoparticles (13–15 nm) have unique properties such as wettability and hardness. Titanates (TiO_2) are used in sunscreens and toothpaste (40–100 nm).[32] Gold NPs have been used in cancer treatment,[33] and zero valent iron NPs have been used as a reducing agent for groundwater remediation.[34]

However, both the size and reactivity of NPs are potential concerns regarding potential health impacts. Nanoparticles have a high surface to volume ratio, which increases reactivity and certain toxicological effects.[36]

8.3.1.3.2 Human Health Effects: Silica Particles. Human exposure to silica may occur at high-levels during the manufacturing process or the application of the nanosol on textiles. The highest potential for exposure occurs during mixing of constituents before polymer synthesis. Because the silica constituents may arrive in bulk and must be poured from the packaging, free silica may be generated, leading to high concentrations of respirable particles. Adequate respiratory protection, ventilation, or "dust control" methods should be used during this step to reduce or eliminate potential exposures. Once applied and cured on the textile, respirable exposures would be minimal unless there is end-of-life disposal that involves bulk mechanical shredding of the textile. Additional exposure to silica can occur during mining and transportation of the bulk silica constituents, preparation of the nanosol, and application of the nanosol onto the textile.

The most toxic form of silica is the free crystalline form (quartz, cristobalite, and tridymite), which causes short, accelerated, and long-term progressive lung injury (silicosis) and carcinogenesis in miners and industrial workers as well as increased risk of rheumatoid arthritis, kidney disease, and tuberculosis.[37] Crystalline silica induces silicosis and pulmonary disease,[37,38] but effects of amorphous silica were not persistent nor fibrogenic. However, animal studies show partially reversible inflammation, granuloma formation, and emphysema but no fibrosis. This lower level of toxicity from amorphous silica NPs was regarded as an effect of more rapid clearance due to lack of surface silanols present in crystalline forms which damage particle-clearing macrophages.[39]

While silica nanosols can be made from either crystalline or amorphous forms of silica, the project team selected the amorphous form because of its low toxicity compared to the crystalline form to account for working environments with lower safety standards. To reduce the risk of respirable silica particles, it was recommended that Gore synthesize the silica nanoparticles directly on the fabric from an organosilane in solution.

Modeling the fate and transport of NPs to organs suggested that nanoparticle translocation to the liver, bone, digestive, and respiratory systems may occur (Figure 8.13); primarily, the NPs enter the lungs and respiratory tract and are then transported into blood and organs.[37] Silica has been

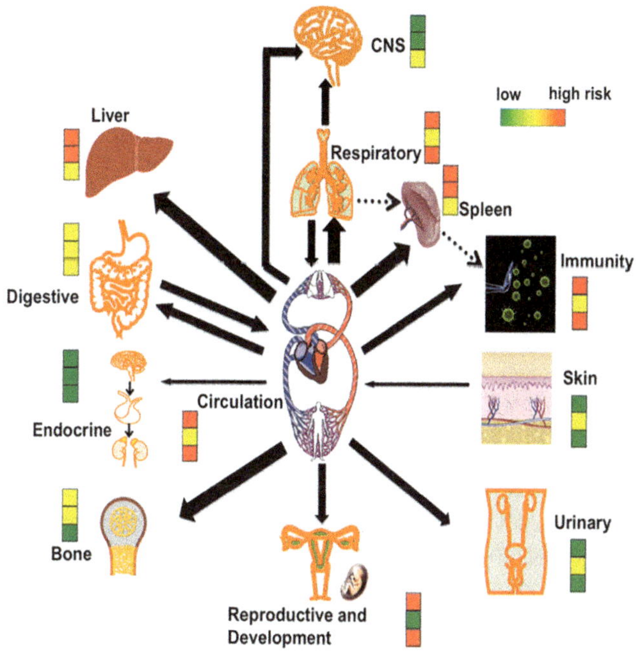

Figure 8.13 Routes of toxicity of nanoparticles with segmented bar showing risk of translocation, repair, and toxicity (top to bottom) in the literature.[40]

shown to exhibit immunosuppressant, hypocholesterolemic, anti-allergenic, anti-apoptotic and anti-histamine functions.[39] Due to many differences between manufacturing practices, size, exposure duration, surface properties, and dose, the human health effects of nanosilica are still uncertain.

8.3.1.3.3 Environmental Toxicity: Silica Particles. Silica is most commonly found in nature as sand or quartz, a ubiquitous material with diverse uses (consumer products, biomedical, drug delivery, and surfacing). Silica NPs from various sources, including road dust, enter the aquatic environment through stormwater runoff and may pose a risk to aquatic organisms.[41] For example, observed toxicity in fish, daphnia, and algae may result from particles that block gills.[42] The presence of different types of NPs had an adverse effect on the microbial diversity and activity in wastewater treatment, and treatment membranes may ultimately be a sink for certain sizes of silica NPs (10–100 nm).[43] Environmental persistence of silica nanosols was assessed to be less significant than the current fluorinated-DWR technology, but data gaps in scientific studies preclude a true comparison of health effects.

A hazard summary of silica nanoparticles is presented in Figure 8.14.

8.3.1.3.4 Hazards of Silanes. The project team's strategy used the silanols APTES and HDTMS to form functional coatings on silica nanomaterials. Both

	Human Toxicity				Environmental Toxicity		
	Mutagenicity	Organ	Respiratory	Reproductive	Persistence	Aquatic	Fate
Silica NPs *Amorphous Silica*	DG	DG	H	DG	L	M	L
APTES	H	H	H	L	M	M	M
HDTMS	DG	M	H	DG	M	M	M

Key	Relative Hazard	High (H)	Moderate (M)	Low (L)	Data Gap (DG)

Figure 8.14 Summary of human and environmental hazards of silica nanosols.[45–57]

of these silanols are considered respiratory toxicants. Thus, providing safe conditions for industrial workers was a primary concern.[44] For instance, manufacturing of APTES requires closed systems with engineering controls (appropriate ventilation, containment, safety equipment). APTES and HDTMS can hydrolyze in water, are biodegradable, and are not persistent.[42] A hazard summary for APTES and HDTMS is presented in Figure 8.14.

The functional coatings of silanes (APTES/HDTMS), both being sensitive to hydrolysis, may require further testing as a consequence of toxicity resulting from degradation products like ethanol and trisalonol. Chronic exposures to nanoparticles may have more unknown health and environmental effects. The functional coatings of silanes (APTES/HDTMS), may need further testing for effects of silanols, a potential degradation product. We did not assess implications of large chemical spills, which may require additional assessments. Skin irritation and respiratory damage from manufacturing processes are important health and safety considerations.

8.3.2 Spinning of Nanofiber Mats

8.3.2.1 Background

The *Jacobaea maritima* (silver ragwort) leaf has an almost superhydrophobic WCA of 147° (see Figure 8.15(A)), achieved due to its hierarchical microstructuring and nanostructuring. The leaf is composed of a tangled net of fibers that are 6 μm in diameter (see Figure 8.15(B)), providing the microstructure. Additionally, each fiber is not smooth but is roughened on the nanoscale (see Figure 8.15(C)).[58]

One way to replicate the hierarchical structuring of the silver ragwort leaf synthetically is by creating a nanofiber mat through a spinning process. Electrospinning can emulate the woven fiber microstructure and even the nanostructured fiber surface (see Figure 8.16).[58,59] Electrospinning uses high voltage to force a charged solution out of a small nozzle. As the solution exits the tip, the solvent evaporates and leaves a polymer fiber with diameters

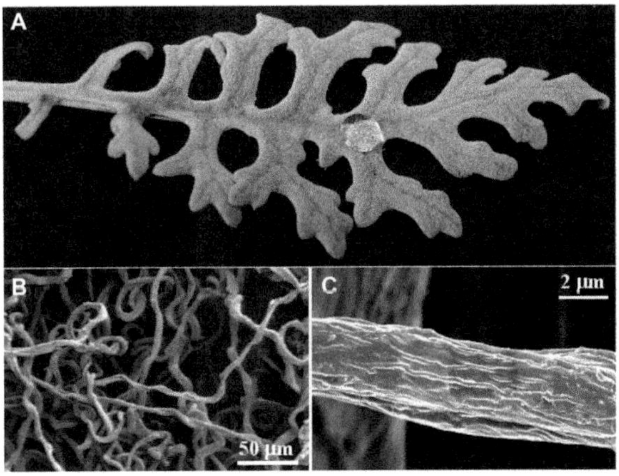

Figure 8.15 Silver ragwort leaf. (A) Digital photograph of a leaf with a beaded water droplet. (B and C) SEM images showing the tangle of fibers and the structuring on the fibers. Reproduced from ref. 58 with permission from Institute of Physics Publishing, Copyright 2006.

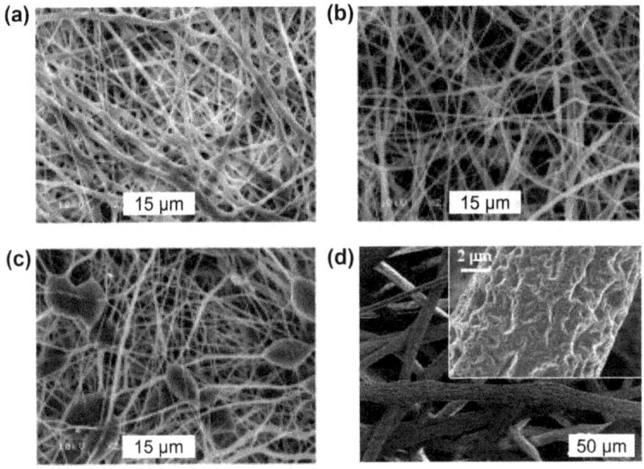

Figure 8.16 SEM examples of fiber mats and fiber morphologies from electrospinning. (a and b) Different fiber diameters and densities from different electrospinning conditions. (c) Fibers with beads and (d) textured fibers are also possible. Images in (a–c) reproduced from ref. 59 with permission from American Chemical Society, Copyright 2005. Image in (d) reproduced from ref. 58 with permission from Institute of Physics Publishing, Copyright 2006.

approximating tens of nanometers. Varying the voltage, solvent, charge, and collection method can result in a variety of fiber diameters, textures, and assemblies (see Figure 8.16).[59,60] Additionally, these fibers can be made from polymers commonly used in garments such as polyamide, PET, and poly(vinyl acetate) (PVA).[60,61]

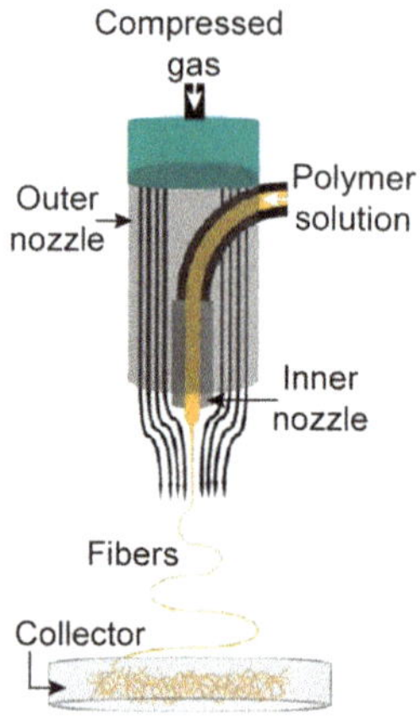

Figure 8.17 Diagram of the solution blow spinning method. High pressure gas shapes a polymer solution into microfibers and nanofibers that are then collected. Reproduced from ref. 62 with permission from American Chemical Society, Copyright 2016.

Instead of using an electric field, solution blow spinning uses an annulus of high pressure gas surrounding the nozzle to create thin polymer fibers (see Figure 8.17). This method is faster and often easier to scale up for industrial applications and can employ a wider range of solvents (including safer solvents). Blow spinning, however, can only produce fibers that are about 100 nm in diameter and have less variability in diameter.[62] With less flexibility in fiber size, the optimal structure for liquid repellency may not be achievable with blow spinning. The project team reviewed more research on hydrophobicity and oleophobicity of electrospun mats to narrow the focus of their discussion.

8.3.2.2 *Performance*

8.3.2.2.1 Hydrophobicity. Highly hydrophobic and superhydrophobic fiber mats can be made using electrospinning. Figure 8.18 shows that while individual fibers have low contact angles, the overall wetting behavior indicates that the fiber mat has a large effective static contact angle due to the hierarchical structure.

The water repellency of the mat can be tuned by carefully selecting solvent(s) and the ratios of solvents and polymers. For example, the static WCA and the water roll-off angle of polystyrene fiber mats vary dramatically with the ratio of two solvents, tetrahydrofuran (THF) and dimethylformamide (DMF) (see Figure 8.19). In this case, a weight ratio of 1 : 3 THF : DMF gave the maximum static WCA and minimum water roll-off angle, which is indicative of good dynamic water repellency (see Figure 8.19).

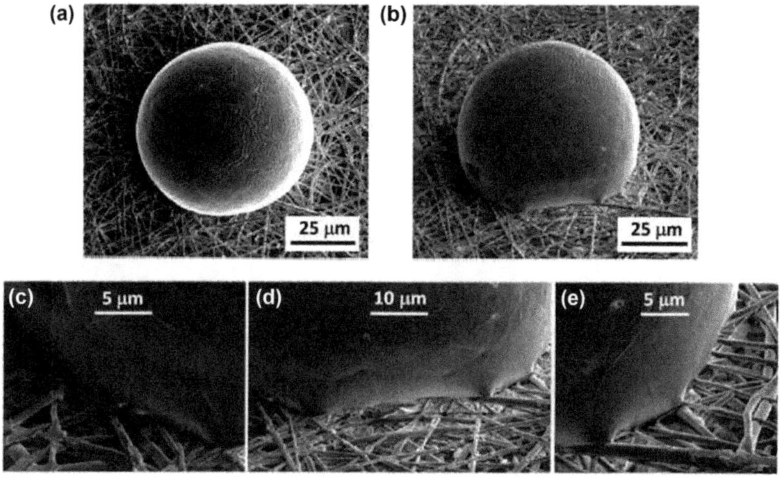

Figure 8.18 SEMs of an oil droplet on an electrospun nanofiber mat. (a and b) Under low magnification, the droplet appears to have a high effective contact angle with the fiber mat, but (c–e) the higher magnifications show the wetting of the individual fibers. Reproduced from ref. 61 with permission from American Chemical Society, Copyright 2015.

These optimal polystyrene fibers had a number of small grooves on the surface, mimicking the silver-ragwort leaf structure. Other studies have shown that fiber mats containing a high density of small polymer beads can also increase hydrophobicity through increased structuring.[62] Sub-micron poly(styrene-*b*-dimethylsiloxane) fiber mats are superhydrophobic with a static WCA of 163° and a contact angle hysteresis of 15°,[63] indicating good static and dynamic water repellency.

8.3.2.2.2 Oleophobicity. Few studies investigated the oleophobicity of non-fluorinated nanoscale fiber mats. Figure 8.20 shows a marked increase in the oil contact angle of a fiber mat compared to a flat surface treated with the same fluorinated chemical. The synthetic fibers such as nylon and polyester are oleophilic; texturing them will not make them oleophobic, so a coating should be used. A featureless DMDMS coating had a diiodomethane contact angle of 71°.[64] Therefore, a textured nanofiber mat coated with DMDS would likely have a greater, oleophobic contact angle.

8.3.2.2.3 Durability, Washability, and Aesthetics. Since nonwoven nanofiber mats are made of the same synthetic fibers that are commonly used in activewear, they should have the same washability, feel, and appearance. Additionally, the fiber mat should not decrease the breathability of

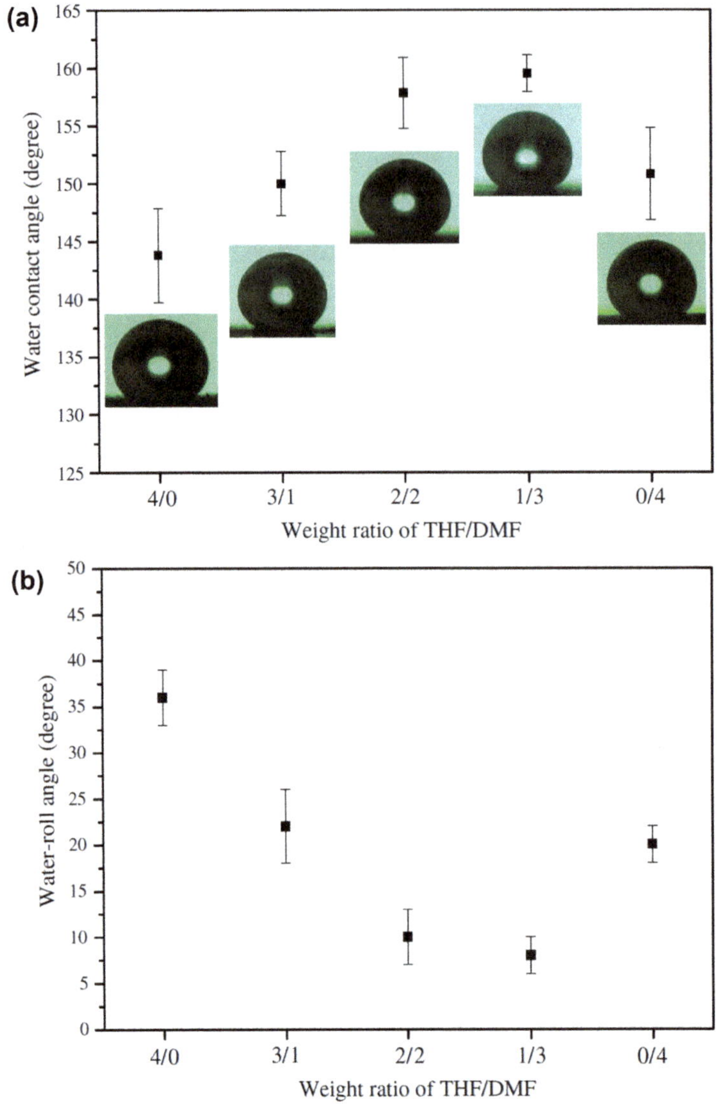

Figure 8.19 Hydrophobicity of electrospun mats. (a) WCA and (b) water roll-off angle for polystyrene fiber mats. By varying the ratio of the two solvents used, tetrahydrofuran (THF) and dimethylformamide (DMF), mats were made with different degrees of hydrophobicity. Reproduced from ref. 58 with permission from Institute of Physics Publishing, Copyright 2006.

the garment.[65] However, there remained uncertainties regarding the durability of fiber mats. It is possible that the three-dimensional web could collapse or be damaged under abrasive forces. It was recommended to use

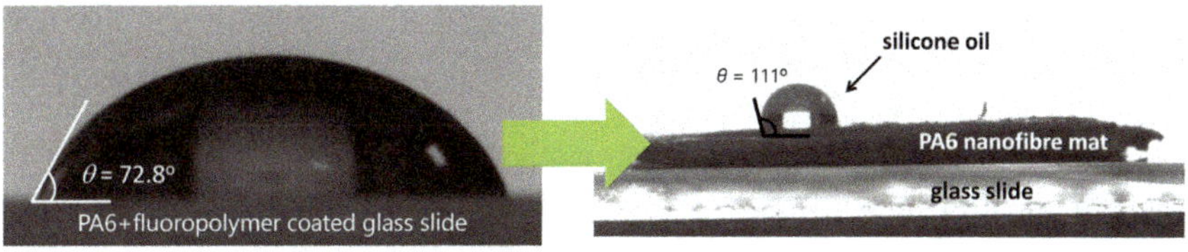

Figure 8.20 Increase in the silicone oil contact angle from a flat coating to a nanofiber mat. The contact angle of the oil on a flat polyamide-6 (nylon) (PA6) surface with a fluoropolymer coating is 72.8° but increases to 111° on a fluoropolymer-coated PA6 fiber mat. Reproduced from ref. 62 with permission from American Chemical Society, Copyright 2016.

the same adhesive for attaching the nanofiber mat as Gore used for attaching the PTFE membrane.

8.3.2.3 Hazards

8.3.2.3.1 Nano and Microfibers. Spun fibers are generally smaller than 100 μm in diameter, posing a risk of microplastic releases during degradation. Microplastics and microfibers are emerging as global contaminants; they are degradation products of plastics and textiles in the size range of 0.1 μm to 5 mm. Each home laundry cycle may produce more than 4500 microfibers; these fibers escape from wastewater treatment plants and contaminate ecosystems.[66] Polyethylene, polypropylene, polystyrene, and nylon are considered significant sources of environmental microfibers. The size of microfibers is similar to that of food for aquatic invertebrates and shellfish, so the fibers are easily ingested. Ingestion of plastics by aquatic life threatens biodiversity, and these plastics may accumulate in the food chain.[67,68] Microplastics and fibers can also absorb high concentrations of lipophilic chemicals.[69]

Carcinogenic compounds such as styrene and benzene are involved in the production of monomers used to make polyester, and these may pose occupational hazards. The actual nylon/polystyrene fibers used in this strategy, however, lead to low occupational harm because polymers are fairly inert. Polyester fibers are not biodegradable. In contrast, fibers made from biodegradable fibers such as polyhydroxyalkanoate (PHA) can degrade to CO_2, methane, and water. Biodegradation occurs through bacterial enzymes, achieving ASTM biodegradation standards of complete removal of carbon in the plastic to CO_2 within 180 days.[70] In a life cycle assessment model, P(3HB) (bioplastic polymer) showed significant reduction in carbon dioxide emissions, greenhouse gases, and fossil fuels compared to polyethylene production.[71]

8.3.2.3.2 Common Spinning Solvents and Greener Alternatives. Many volatile organic solvents are flammable, acute and chronic toxicants for humans, and environmental hazards. Fiber spinning involves the selection of common solvents: THF, DMF, acetone, and dimethyl sulfoxide (DMSO). Selecting a safer solvent was key to minimizing the hazard associated with these fiber mats. Greener alternatives such as acetone, cyclopentylmethylether (CPME), ethanol, or water produce fewer peroxides than THF and have less solvent loss during reactions.[72–76] CPME can be produced in a nontoxic method, is more stable than THF, resists peroxide formation, and improves laboratory safety. Ethyl lactate is a commercially available (Sigma) safer alternative to acetone, is not a carcinogen, and is easily recyclable.[76] The hazards of different solvents are summarized in Figure 8.21.

8.3.2.3.3 Solvent Exposure Control and Considerations.

Potential exposure windows occur throughout the life cycle; during synthesis of the solvents and their constituents, preparation of the solution, formation of the fiber mat, application to the textile, and curing. Once applied and cured on the textile, respirable exposures to consumers should be non-existent as the volatile solvents evaporate during the curing process. The population of concern for exposure to high levels of volatile solvents consists of occupational workers in the manufacturing of the textile. Solvent constituents arrive in bulk and must be poured from the containers in which they were shipped into a hopper for mixing. As a result, the solvents may generate harmful vapors at high concentrations as well as pose a risk of dermal exposure. Adequate respiratory protection, dermal protection, eye protection, ventilation, or engineering controls should be used during these steps to reduce or eliminate potential exposures.

Spinning fiber mats to produce DWR materials affords the possibility of producing greener textiles and using greener solvents. The traditional use of

	Human Toxicity				Environmental Toxicity		
	Mutagenicity	Organ	Respiratory	Reproductive	Persistence	Aquatic	Fate
THF	H	H	H	L	L	L	L
DMF	H	H	H	H	L	L	L
DMSO	M	M	M	H	L	DG	L
Acetone	M	H	H	DG	L	L	L
CPME	L	M	M	L	L	L	L

Key	Relative Hazard	High (H)	Moderate (M)	Low (L)	Data Gap (DG)

Figure 8.21 Relative comparison of the human and environmental hazards of the solvents used in spinning.[57,77–100]

solvents such as THF/DMF, acetone, and butanol may not be as environmentally problematic as fluorinated DWR textile production. However, these solvents could be replaced by ethyl lactate or CPME.

8.4 Conclusions

8.4.1 Hazard Comparison

Figure 8.22 compares the hazard traits of the two most promising alternatives with those of traditional fluorinated DWR. This figure was based on relative hazard; low hazard (green) did not necessarily mean "safe." All compounds can be poisonous;[101,102] therefore potency, exposure, dose, and population sensitivity would have to be considered for each endpoint, even when a chemical is "greener."

Figure 8.22 first lists the "bad actors" of this chapter, the PFASs. PFOA was the baseline against which the other chemicals were compared. Overall, PFHxA was less hazardous than PFOA but included more data gaps. Both were persistent and may bioaccumulate, increasing their environmental toxicity.

The next rows, silica NPs, APTES, and HDTMS, correspond to the silica nanosol alternative. This set of chemicals had several hazard data gaps, and additionally had medium to high hazards in both human and environmental toxicity. The project team found that silica nanoparticles had some of the highest hazards due to their high potential aquatic toxicity.

The last rows include solvents reported in the electrospinning literature (THF and DMF) which adversely affect the human health, but pose a

Alternative	Human Toxicity				Environmental Toxicity		
	Mutagenicity	Organ	Respiratory	Reproductive	Persistence	Aquatic	Fate
PFOA	H	H	L	H	H	H	H
PFHxA	L	DG	DG	DG	M	M	H
Silica NPs	L	DG	H	DG	L	H	L
APTES	DG	M	H	L	M	M	L
HDTMS	DG	M	H	L	M	M	L
THF	H	M	H	H	L	L	L
DMF	H	H	M	H	L	DG	L
DMSO	M	M	M	H	L	L	L
Acetone	L	M	M	H	L	L	L
CPME	L	M	M	L	L	L	L

Key

Relative Hazard	High (H)	Moderate (M)	Low (L)	Data Gap (DG)
Alternative	Contains PFASs	Nanosols	Spinning	

Figure 8.22 Hazard table comparing currently-used DWR materials to our alternatives.

relatively low environmental hazard. The proposed safer solvents (DMSO, acetone, and CPME) exhibit a range of human hazards including cell toxicity and reproductive toxicity. Blow spinning could likely offer more flexibility in the solvents used and, therefore, safer solvents could be chosen over THF and DMF. Polymers used to make the fibers are not included in the table since many of them were already in use in the textile industry.

8.4.2 Performance Comparison

A spider diagram was constructed to compare the performance of a PFAS-based DWR with silica nanosols and electrospun fiber mats, with endpoints of human and environmental hazards and performance metrics (see Figure 8.23). PFAS performance served as the comparison, with its hydrophobicity, oleophobicity, durability, and aesthetics (feel, color, and breathability) being the model for performance. As shown in Figure 8.23, if the line touched the outer ring, it signified a high rating; the larger the area contained in the curve, the better it performed.

Both nanosols and electrospun fiber mats achieved hydrophobicity comparable to that of PFASs, but both fell short in oleophobicity and durability metrics. Additionally, spinning resulted in excellent breathability, feel, and color, but nanosols had shortcomings in aesthetics. When the performance metrics were combined with hazard assessment, spinning out-performed nanosols and PFASs, except for durability and oleophobicity. Spinning achieved the highest marks for environmental toxicity and fate. Based on these results and the hazard assessment, the project team identified

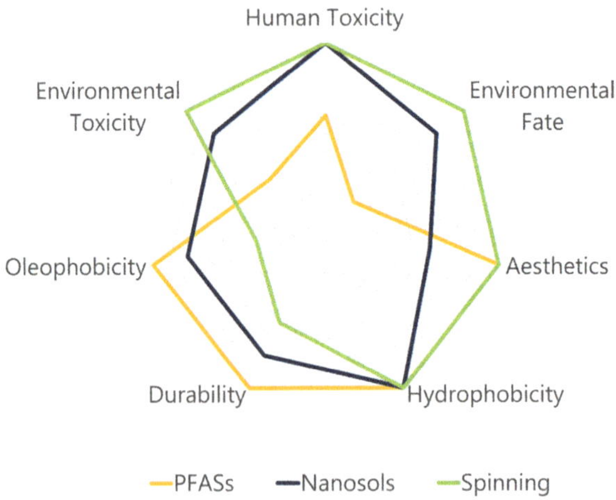

Figure 8.23 Relative hazard and performance comparison between PFAS and the two alternatives presented in this chapter. A strategy is better performing if its endpoints lie closer to the outer ring of the chart. More hazardous and poorer performing alternatives will score closer to the center.

spinning as the most promising solution for Gore to continue researching. The project team also recommended looking at oleophobic modifications of the electrospun fiber mats, such as a dimethyldimethoxysilane coating.

8.4.3 Feasibility

Finally, the project team assessed whether the 'greener solutions' had been successfully tested on fabrics and if Gore used similar processes and materials in fabric production. Based on these criteria, rankings were provided on a scale of one to five, five being the most feasible (already used in industry with success).

Silica nanosols had been tested on fabrics and had been shown to produce highly hydrophobic surfaces on polyester. However, the literature at the time did not demonstrate durability of silica nanosols through rigorous wash–dry cycles, and using dimethyldimethoxysilane to improve oleophobicity had not been tested on fabrics. Silica nanosols could be applied using the common dip and nip application method, and only required three to four compounds for the solution. All materials needed (amorphous silica and various silanes) were commercially available from vendors such as Sigma Aldrich. Because further testing would have been required for the washability, durability, and oleophobicity of nanosol coated fabrics, the feasibility of silica nanosols was given a rank of three.

The spinning process creates microfiber mats out of polymers to create textured, tangled fibers, which have direct applicability to fabric production. These mats were shown to be superhydrophobic, breathable, and washable, but the washability had not been thoroughly investigated, nor had the oleophobicity of these mats been tested. This process could potentially use only a gas, one or two solvents, the desired polymer (*e.g.*, polystyrene), and a binding agent. However, binding this textured mat on top of the base Gore fabric had not been tested, and further testing of the oleophobicity was needed. Thus, the feasibility of spinning was given a rank of four.

8.4.4 Next Steps

The project team recommended optimizing blow spinning nonwoven nanofiber mats with safer solvents (water, ethanol, acetone, *etc.*) and exploring the use of biodegradable polymers if the fate of microplastics aligned with Gore's long-term lifecycle goals. The solvent(s) used and experimental conditions (the type of gas solvent, polymer, solution concentration, and amount of pressure) should be modified to produce high-performing materials. Particular attention should be paid to the oleophobicity of the material; experiments with dimethyldimethoxysilane coatings should be conducted to determine if the coating imparts the desired oleophobicity to the final fabric product. It was recommended that Gore first experiment by using the same adhesion technique to attach the nanofiber mat that the company used for attaching the PTFE membrane to determine feasibility.

For pursuing nanosol coating as the DWR layer on fabrics, the project team recommended to first test a silica nanosol solution (silanes, organosilanes, SiO_2, DMDMS, and water) on Gore's fabrics to determine the baseline performance. If the nanosol performed well, the team recommended varying the concentrations of the solution to determine which concentrations do not affect breathability, flexibility, or dying processes. Once hydrophobicity and oleophobicity were attained, Gore could test the fabric using their standard durability and washability tests. A combination of a nanosol coating on a nanofiber mat can also be explored.

8.4.5 Recommendations for Non-fluorinated DWR

Considering the proven performance, known hazards, data gaps, and feasibility of the two non-fluorinated DWR alternatives presented here, a spun nanofiber mat with a DMDMS coating was the most promising alternative. An illustration of this solution is depicted in Figure 8.24. Spinning creates a hierarchical microstructure and nanostructure, which the silver ragwort leaf and other animal and plant species use to repel liquids. The addition of a DMDMS coating is necessary to impart oil repellency to the structure, and the texture of the fiber mat would improve the baseline oleophobicity of a smooth DMDMS coating. Due to the covalent bonding of DMDMS to the fiber mat, this coating was expected to be durable. However, spun fiber mats were not subjected to rigorous abrasion and washing tests necessary to ensure that the nanofibers would withstand normal wear and use. The nanofibers were composed of polymers already used in wearable textiles and the open structure imparted more breathability than a traditional weave. Additionally, it was expected that DMDMS would not impact the aesthetics because of its chemical similarity to the silicone nanofibers, without detracting from the appearance of the fabric.

8.5 State of Gore DWR Materials Present Day

Since the time of writing this report, Gore Fabrics Division has achieved the goal of producing materials that are free of PFCs of environmental concern over the lifecycle of its consumer products. These initiatives are summarized

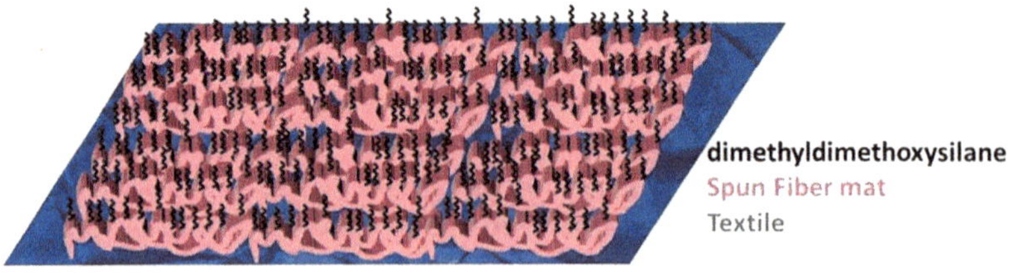

Figure 8.24 Graphical depiction of the most promising strategy. A spun fiber mat coated with DMDMS is used as the DWR coating on the garment textile.

in the 'Sustainability Framework of Gore Fabrics' and include increasing product life and recyclability, and developing DWR materials with lower environmental footprints.[103]

Although fluorinated compounds such as PFOA were eliminated from Gore fabrics in 2013, fluoropolymers are still used in GORE-TEX fabrics containing a thin membrane called ePTFE. According to Gore, the 'science of fluoromaterials' is gaining acceptance from the fabric industry as a result of growing evidence of PTFE's low environmental hazard due to its reduced transformation in the environment.[104]

As of 2022, Gore launched expanded polyethylene (ePE) as a new base material for their non-fluorinated membrane technologies. These materials combined with polyurethane, create a waterproof, windproof, and breathable membrane; however, it is unclear how ePE performs with oleophobicity metrics. Due to an ever increasing demand for durable, water-repellent, oil-repellent fabrics, research and initiatives for environmentally safer, non-fluorinated solutions will continue for many years.

Abbreviations

AATCC	American Association of Textile Chemists and Colorists
APTES	3-Aminopropyltriethoxysilane
BCF	Bioconcentration factor
CA	Contact angle
CAH	Contact angle hysteresis
CPME	Cyclopentylmethylether
DMDMS	Dimethyldimethoxysilane
DMF	Dimethylformamide
DMSO	Dimethyl sulfoxide
DWR	Durable water repellency
ECHA	European Chemicals Agency
EPA	United States Environmental Protection Agency
Gore	W. L. Gore & Associates
GPTMS	3-Glycidoxypropyltrimethoxysilane
HDTMS	Hexadecyltrimethoxysilane
K_{ow}	Octanol–water partitioning coefficient
NP	Nanoparticle
P(3HB)	Poly(3-hydroxybutyrate)
PA6	Polyamide-6
PBT	Persistent, bioaccumulative, and toxic
PDMS	Poly(dimethylsiloxane)
PET	Poly(ethylene terephthalate)
PFASs	Per- and polyfluoroalkyl substances
PFC-EC	Per- and polyfluorinated chemicals of environmental concern
PFHxA	Perfluorohexanoic acid
PFOA	Perfluorooctanoic acid
PHA	Polyhydroxyalkanoate

PhTES	Phenyltriethoxysilane
PMSQ	Polymethylsilsesquioxane
PTFE	Poly(tetrafluoroethylene)
PVA	Poly(vinyl acetate)
REACH	European Union's Registration, Evaluation, Authorisation and Restriction of Chemicals
SEM	Scanning electron microscope
SiO_2	Silica
SLIPS	Slippery liquid infused porous structures
TEOS	Tetraethoxysilane
THF	Tetrahydrofuran
TiO_2	Titania
UV	Ultraviolet
WCA	Water contact angle

Acknowledgements

We are grateful for the opportunity to publish our 'greener solutions' approaches as a result of the Greener Solutions course directed by Dr Tom McKeag, Dr Meg Schwarzman, and Dr Marty Mulvihill at UC Berkeley. We thank Dr Barbara Henry, partner in the Greener Solutions challenge issued by W. L. Gore, and mentor for the Greener Solutions course. For helpful discussions about oleophobic textiles, we recognize Dr Robert C. Buck. Finally, this challenge would not have been possible without Dr. Robert Gore, an original founder of the DWR fabrics industry.

References

1. C. D. Vecitis, H. Park, J. Cheng, B. T. Mader and M. R. Hoffmann, *Front. Environ. Sci. Eng. China*, 2009, **3**, 129–151.
2. T. A. Bruton and A. Blum, *Environ. Health*, 2017, **16**(1), 120.
3. S. K. Ritter, *Chem. Eng. News*, 2010, **88**(5), 12–17.
4. Z. Wang, I. T. Cousins, M. Scheringer and K. Hungerbühler, *Environ. Int.*, 2013, **60**, 242–248.
5. Greenpeace PFC Revolution in the Outdoor Sector, https://www.greenpeace.org/static/planet4-international-stateless/2017/02/432f4056-pfc-revolution-in-outdoor-sector.pdf (accessed April 2023).
6. ECHA-17-G-09-EN Cat. Number: ED-01-17-291, Annex XIII of REACH. Guidance on Information Requirements and Chemical Safety Assessment Part C: PBT/vPvB Assessment Reference: ECHA-17-G-09-EN Cat. Number: ED-01-17-291, https://reachonline.eu/reach/en/annex-xiii-3-3.2.html (accessed April 2023).
7. T. Darmanin and F. Guittard, *Mater. Today*, 2015, **18**(5), 273–285.
8. T. L. Liu and C. J. C. Kim, *Science*, 2014, **346**(6213), 1096–1100.
9. S. Michielsen and H. J. Lee, *Langmuir*, 2007, **23**, 6004–6010.

10. Environmentally Friendly Water and Dirt Repellence, https://www.schoeller-textiles.com/en/technologies/ecorepel (accessed April 2023).
11. T. Kim, H. Kang and N. Yoon, *Fibers Polym.*, 2017, **18**, 285.
12. L. Gao and T. J. McCarthy, *Langmuir*, 2006, **22**, 5998–6000.
13. J. Zimmermann, F. A. Reifler, G. Fortunato, L. C. Gerhardt and S. Seeger, *Adv. Funct. Mater.*, 2008, **18**, 3662–3669.
14. P. Kim, M. J. Kreder, J. Alvarenga and J. Aizenberg, *Nano Lett.*, 2013, **13**, 1793.
15. Q. Liu, Y. Yang, M. Huang, Y. Zhou, Y. Liu and X. Liang, *Appl. Surf. Sci.*, 2015, **346**, 68.
16. D. F. Cheng, C. Urata, M. Yagihashi and A. Hozumi, *Angew. Chem., Int. Ed.*, 2012, **51**, 2956.
17. V. G. Damle, A. Tummala, S. Chandrashekar, C. Kido, A. Roopesh, X. Sun, K. Doudrick, J. Chinn, J. R. Lee, T. P. Burgin and K. Rykaczewski, *ACS Appl. Mater. Interfaces*, 2015, **7**, 4224.
18. N. A. Ibrahim, R. Refaie and A. F. Ahmed, *J. Ind. Text.*, 2010, **40**, 65.
19. A. Anjum, M. Zuber, K. M. Zia, A. Noreen, M. N. Anjum and S. Tabasum, *Int. J. Biol. Macromol.*, 2016, **89**, 161.
20. M. Sadat-Shojai, *J. Mater. Sci. Technol.*, 2016, **32**, 1013.
21. V. Urtuvia, P. Villegas, M. González and M. Seeger, *Int. J. Biol. Macromol.*, 2014, **70**, 208.
22. E. Ten, D. F. Bahr, B. Li, L. Jiang and M. P. Wolcott, *Ind. Eng. Chem. Res.*, 2012, **51**, 2941.
23. Q. Zhao, L. Y. L. Wu, H. Huang and Y. Liu, *Mater. Des.*, 2016, **92**, 541.
24. W. A. Daoud, J. H. Xin and X. Tao, *J. Am. Ceram. Soc.*, 2004, **87**, 1782.
25. H. Wang, J. Ding, Y. Xue, X. Wang and T. Lin, *J. Mater. Res.*, 2010, **25**, 1336.
26. M. Yu, G. Gu, W. D. Meng and F. L. Qing, *Appl. Surf. Sci.*, 2007, **253**, 3669.
27. M. Ardanuy, M. Faccini, D. Amantia, L. Aubouy and G. Borja, *Surf. Coat. Technol.*, 2014, **239**, 132.
28. B. Mahltig and H. Böttcher, *J. Sol–Gel Sci. Technol.*, 2003, **27**, 43.
29. L. Wang and T. J. McCarthy, *Angew. Chem., Int. Ed.*, 2016, **128**, 252.
30. B. Mahltig and T. Textor, *Nanosols and textiles*, World Scientific, Singapore, 2008.
31. M. Sajid, M. Ilyas, C. Basheer, M. Tariq, M. Daud, N. Baig and F. Shehzad, *Environ. Sci. Pollut. Res.*, 2015, **22**, 4122.
32. A. L. Doiron, B. Clark and K. D. Rinker, *Biotechnol. Bioeng.*, 2011, **108**, 2988.
33. P. Ghosh, G. Han, M. De, C. K. Kim and V. M. Rotello, *Adv. Drug Delivery Rev.*, 2008, **60**, 1307.
34. A. N. Bezbaruah, S. Krajangpan, B. J. Chisholm, E. Khan and J. J. Elorza Bermudez, *J. Hazard. Mater.*, 2009, **166**, 1339.
35. Nanotechnology definition, https://chembam.com/definitions/nanotechnology/ (accessed April 2023).
36. A. Nel, T. Xia, L. Mädler and N. Li, *Science*, 2006, **311**, 622.
37. Y. Zhang, Y. Bai, J. Jia, N. Gao, Y. Li, R. Zhang, G. Jiang and B. Yan, *Chem. Soc. Rev.*, 2014, **43**, 3762.
38. C. C. Leung, I. T. S. Yu and W. Chen, *Lancet*, 2012, **379**(9830), 2008–2018.

39. S. Murugadoss, D. Lison, L. Godderis, S. Van Den Brule, J. Mast, F. Brassinne, N. Sebaihi and P. H. Hoet, *Arch. Toxicol.*, 2017, **91**, 2967.
40. H. Buckley, *The Conversation*, 2015, https://theconversation.com/safer-chemicals-would-benefit-both-consumers-and-workers-47445 (accessed April 2023).
41. Y. Yang, M. Vance, F. Tou, A. Tiwari, M. Liu and M. F. Hochella, *Environ. Sci.: Nano*, 2016, **3**, 534.
42. PubChem 3-Aminopropyltriethoxysilane Compound Summary, https://pubchem.ncbi.nlm.nih.gov/compound/3-Aminopropyltriethoxysilane (accessed April 2023).
43. M. C. Roco, *J. Nanopart. Res.*, 2011, **13**, 427.
44. A. L. Quinn, J. M. Regan, J. M. Tobin, B. J. Marinik, J. M. McMahon, D. A. McNett, C. M. Sushynski, S. D. Crofoot, P. A. Jean and K. P. Plotzke, *Toxicol. Sci*, 2007, **96**, 145.
45. V. Rabolli, L. C. J. Thomassen, F. Uwambayinema, J. A. Martens and D. Lison, *Toxicol. Lett.*, 2011, **206**, 197.
46. E. Maser, M. Schulz, U. G. Sauer, M. Wiemann, L. Ma-Hock, W. Wohlleben, A. Hartwig and R. Landsiedel, *Mutat. Res., Genet. Toxicol. Environ. Mutagen.*, 2015, **794**, 57.
47. C. Uboldi, G. Giudetti, F. Broggi, D. Gilliland, J. Ponti and F. Rossi, *Mutat. Res., Genet. Toxicol. Environ. Mutagen.*, 2012, **745**, 11.
48. H. Ernst, S. Rittinghausen, W. Bartsch, O. Creutzenberg, C. Dasenbrock, B. D. Görlitz, M. Hecht, U. Kairies, H. Muhle, M. Müller, U. Heinrich and F. Pott, *Exp. Toxicol. Pathol.*, 2002, **54**, 109.
49. C. Marquardt, S. Fritsch-Decker, M. Al-Rawi, S. Diabaté and C. Weiss, *Toxicology*, 2017, **379**, 40.
50. M. Tsugita, N. Morimoto and M. Nakayama, *Part. Fibre Toxicol.*, 2017, **14**, 11.
51. W. Miao, B. Zhu, X. Xiao, Y. Li, N. B. Dirbaba, B. Zhou and H. Wu, *Aquat. Toxicol.*, 2015, **161**, 117.
52. H. Xie and J. Wu, *Chem. – Biol. Interact.*, 2016, **258**, 197.
53. Project Pharos chemical and material online catalog of chemicals and GreenScreen scores, https://pharosproject.net/ (accessed April 2023).
54. ECHA/REACH Silicon Dioxide Dossier, https://echa.europa.eu/registration-dossier/-/registered-dossier/15556 (accessed April 2023).
55. ECHA/REACH 3-aminopropyltriethoxysilane Dossier, https://echa.europa.eu/registration-dossier/-/registered-dossier/15027 (accessed April 2023).
56. S. Wang, Z. Liu, W. Wang and H. You, *RSC Adv.*, 2017, **7**, 37065.
57. PubChem Hexadecyltrimethoxysilane Compound Summary, https://pubchem.ncbi.nlm.nih.gov/compound/Hexadecyltrimethoxysilane (accessed April 2023).
58. Y. Miyauchi, B. Ding and S. Shiratori, *Nanotechnology*, 2006, **17**, 5151.
59. M. Ma, Y. Mao, M. Gupta, K. K. Gleason and G. C. Rutledge, *Macromolecules*, 2005, **38**, 9742.
60. W. E. Teo and S. Ramakrishna, *Nanotechnology*, 2006, **17**, R89.

61. U. Stachewicz, R. J. Bailey, H. Zhang, C. A. Stone, C. R. Willis and A. H. Barber, *ACS Appl. Mater. Interfaces*, 2015, **7**, 16625.
62. J. L. Daristotle, A. M. Behrens, A. D. Sandler and P. Kofinas, *ACS Appl. Mater. Interfaces*, 2016, **8**, 34951.
63. M. Ma, R. M. Hill, J. L. Lowery, S. V. Fridrikh and G. C. Rutledge, *Langmuir*, 2005, **21**, 5549.
64. L. Wang and T. J. McCarthy, *Angew. Chem., Int. Ed.*, 2016, **1281**, 252.
65. S. Lee and S. K. Obendorf, *J. Appl. Polym. Sci.*, 2006, **102**, 3430.
66. M. Cole, *Sci. Rep.*, 2016, **6**, 34519.
67. J. A. Ivar Do Sul and M. F. Costa, *Environ. Pollut.*, 2014, **185**, 352.
68. C. M. Rochman, A. M. Cook and A. A. Koelmans, *Environ. Toxicol. Chem.*, 2016, **35**, 1617.
69. H. Lee, W. J. Shim and J. H. Kwon, *Sci. Total Environ.*, 2014, **470–471**, 1545.
70. PLA and PHA Biodegradation in the Marine Environment, https://www2.calrecycle.ca.gov/Publications/Download/1006?opt=dln (accessed April 2023).
71. S. Balaji, K. Gopi and B. Muthuvelan, *Algal Res.*, 2013, **2**, 278.
72. S. Aparicio and R. Alcalde, *Green Chem.*, 2009, **11**, 65.
73. I. T. Horváth, H. Mehdi, V. Fábos, L. Boda and L. T. Mika, *Green Chem.*, 2008, **10**, 238.
74. D. F. Aycock, *Org. Process Res. Dev.*, 2007, **11**, 156.
75. K. Watanabe, N. Yamagiwa and Y. Torisawa, *Org. Process Res. Dev.*, 2007, **11**, 251.
76. Sigma Aldrich Greener Solvent Alternatives: Supporting the Advancement of Chemistry through Sound Environmental, Social & Fiscal Responsibilities, https://www.sigmaaldrich.com/deepweb/assets/sigmaaldrich/product/documents/115/677/greener_solvent_alternatives.pdf (accessed April 2023).
77. Pharos Project, Online catalog of chemicals and GreenScreen scores, 2017, https://www.pharosproject.net/material/chemical.
78. C. J. Choi, E. K. Rushton, A. Vardy, L. Higgins, A. Augello and R. J. Parod, *Toxicol. Lett.*, 2017, **276**, 138.
79. ECHA/REACH Tetrahydrofuran Dossier, https://echa.europa.eu/registration-dossier/-/registered-dossier/15474/7/3/2 (accessed April 2023).
80. M. Corvaro, S. Gehen, K. Andrews, R. Chatfield, F. Macleod and J. Mehta, *Regul. Toxicol. Pharmacol.*, 2017, **89**, 131.
81. PubChem Tetrahydrofuran Compound Summary, https://pubchem.ncbi.nlm.nih.gov/compound/8028 (accessed April 2023).
82. The Comparative Toxicogenomics Database (CTD), http://ctdbase.org/ (accessed April 2023).
83. Z. Lv, Y. Yao, Z. Lv, L. Sang and H. Min, *Toxicol. Environ. Chem.*, 2008, **90**(6), 1177–1186.
84. Toxicological Review of Tetrahydrofuran (CAS 109-99-9) in Support of Summary Information on the Integrated Risk Information System (IRIS), https://iris.epa.gov/static/pdfs/1023tr.pdf (accessed April 2023).

85. J. Walrath, W. E. Fayerweather, P. G. Gilby and S. Pell, *J. Occup. Med.*, 1989, **31**, 432.
86. K. I. Tanaka and T. Utsunomiya, *Jpn. J. Ind. Health*, 1982, **24**, 3.
87. ECHA/REACH *N*,*N*-dimethylformamide Substance Infocard, https://echa.europa.eu/substance-information/-/substanceinfo/100.000.617 (accessed April 2023).
88. H. U. Käfferlein, T. Göen, J. Müller, R. Wrbitzky and J. Angerer, *Int. Arch. Occup. Environ. Health*, 2000, **73**, 113.
89. H. Miyauchi, Y. Tsuda, A. Minozoe, S. Tanaka, H. Arito, T. Tsukahara and T. Nomiyama, *Ind. Health*, 2014, **52**, 512.
90. J. Galvao, B. Davis, M. Tilley, E. Normando, M. R. Duchen and M. F. Cordeiro, *FASEB J.*, 2014, **28**, 1317.
91. S. W. Jacob and J. C. de la Torre, *Pharmacol. Rep.*, 2009, **61**, 225.
92. J. L. Hanslick, K. Lau, K. K. Noguchi, J. W. Olney, C. F. Zorumski, S. Mennerick and N. B. Farber, *Neurobiol. Dis.*, 2009, **34**, 1.
93. L. Penazzi, J. Lorengel, F. Sündermann, N. Golovyashkina, S. Marre, C. M. B. Mathis, L. Lewejohann, R. Brandt and L. Bakota, *Neuropharmacology*, 2017, **113**, 434.
94. K. Takeda, M. Pokorski, Y. Sato, Y. Oyamada and Y. Okada, *Adv. Exp. Med. Biol.*, 2016, **885**, 89.
95. Evidence on the Carcinogenicity of *N*,*N*-dimethylformamide, https://oehha.ca.gov/media/downloads/proposition-65/chemicals/dmfhid080808.pdf (accessed April 2023).
96. R. M. David, H. S. Jones, G. H. Panter, M. J. Winter, T. H. Hutchinson and J. K. Chipman, *Chemosphere*, 2012, **88**, 912.
97. A. Farkas, L. Kemény, M. Széll, A. Dobozy and Z. Bata-Csörgo, *Arch. Dermatol. Res.*, 2003, **295**, 56.
98. ECHA/REACH Acetone Dossier, https://echa.europa.eu/registration-dossier/-/registered-dossier/15460/7/6/4 (accessed April 2023).
99. K. Watanabe, *Molecules*, 2013, **18**(3), 3183–3194.
100. ECHA/REACH Cyclopentane, methoxy-Dossier, https://echa.europa.eu/registration-dossier/-/registered-dossier/26626 (accessed April 2023).
101. J. F. Borzelleca, *Toxicol. Sci.*, 2000, **53**, 2.
102. D. J. Farrell and L. Bower, *J. Clin. Pathol.*, 2003, **56**, 803.
103. Gore.com, Gore Fabrics Division Responsibility Update, 2021, https://www.gore.com/about/responsible-enterprise (accessed May 2023).
104. B. Henry, *Gore.com*, Summary of PTFE Studies to Investigate Persistence, Degradation, Transformation to or Release of Substances of Concern, 2022, https://www.gore.com/resources/summary-of-ptfe-studies (accessed May 2023).

CHAPTER 9

PFAS-free Moisture Barriers in Structural Firefighting Gear

G. E. CAMPBELL,[a] S. GLAZER,[a] B. STINGER,*[b] M. THOMPSON[a] AND S. THOMPSON[a]

[a] University of California, Berkeley, School of Public Health, Berkeley, CA, 94704, USA; [b] Department of Chemistry, University of California, Berkeley, CA, 94720, USA
*Email: bstinger@berkeley.edu

9.1 Challenge Overview

The composition of firefighter turnout gear has come under scrutiny due to the adverse health effects of perfluoroalkyl and polyfluoroalkyl substances (PFASs) found in the gear.[1] PFASs have been linked to certain cancers, thyroid disease, reproductive problems, and other serious adverse health effects.[2,3] Firefighters are exposed to many carcinogenic chemicals through smoke and firefighting materials, such as aqueous film forming foams (AFFF), discussed in Chapter 1. Finding alternatives to PFASs in firefighting turnout gear needs to be prioritized to reduce cumulative occupational exposure for firefighters.

Existing firefighter turnout gear is made up of three layers: an inner thermal layer closest to the skin, a middle moisture barrier, and a durable water repellent (DWR) outer shell (see Figure 9.1). The primary function of the moisture barrier is to block penetration of water, liquid chemicals, and heat, while also allowing perspiration and body heat to escape.[1] Liquid resistance and vapor permeability are particularly important for the prevention of steam burns, which occur if water and heat get trapped inside the gear.[4,5]

Green Chemistry Series No. 81
Toward a PFAS-free Future: Safer Alternatives to Forever Chemicals
Edited by Simona A. Bălan, Thomas A. Bruton and Kimberly G. Hazard

Published by the Royal Society of Chemistry, www.rsc.org

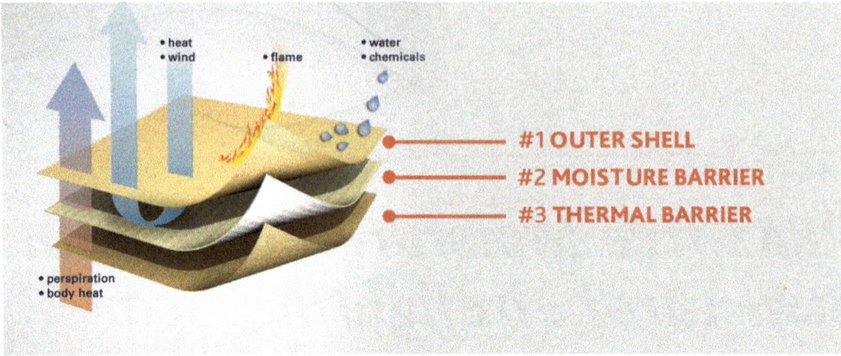

Figure 9.1 Schematic showing the three layers of firefighter turnout gear. Reproduced from ref. 81.

The National Fire Protection Association (NFPA) sets standards on Protective Ensembles for Structural Fire Fighting and Proximity Fire Fighting in the United States and Canada. The currently used moisture barrier layers, approved by the NFPA 1971 Standards, contain a polytetrafluorethylene (PTFE) laminate. PTFE, a synthetic fluoropolymer, is one of the most well-known and commonly applied PFASs. It is effective in turnout gear due to its water and oil resistance, thermal stability, and breathability.[1] PTFE itself is considered inert; however, other PFASs used in the PTFE manufacturing process are present on the gear and pose health risks to the wearer. Studies found that PFASs migrate across turnout gear layers into untreated material such as the inner thermal layer.[6] Elevated blood levels of PFASs have been measured in firefighters, and dermal absorption through direct skin contact with turnout gear is one key exposure route.[1,7,8]

Firefighting as an occupational exposure is classified as an IARC group 1 known human carcinogen.[2] The National Institute for Occupational Safety and Health (NIOSH) reports that firefighters have a 9% higher risk of cancer diagnosis and a 14% higher risk of cancer-caused mortality than the total U.S. population.[9] It is also important to consider the healthy worker effect, which suggests that workers are typically healthier than the general population due to their ability to work. Firefighters are especially likely to be healthier than the general population since their work includes strenuous physical activity, which suggests that these statistics may underrepresent the true incidence of cancer in them.[10]

Each turnout gear ensemble has a lifespan of ten years or less depending on the condition, as required by NFPA 1971.[11] Promptly finding safer alternatives to PFASs in this gear that could be worn for up to a decade is important to prevent adverse health outcomes in firefighters and future harm to the environment.

This project comprised a semester-long investigation of alternatives to PTFE in firefighting turnout gear as part of the UC Berkeley graduate course, Greener Solutions, from August to December 2022. Course instructors

(Dr Megan Schwarzman, Dr William Hart-Cooper, Sophia Steffens, and Kimberly Hazard) facilitated and guided the partnership between Derek Urwin, Assistant Adjunct Professor of Chemistry and Biochemistry at UCLA, Neil McMillan, Director of Science and Research at the International Association of Firefighters (IAFF), and the student team consisted of Brittany Stinger (chemistry), Grace Campbell (environmental health sciences), McKenna Thompson (environmental health sciences), Sophia Glazer (environmental health sciences) and Sophia Thompson (environmental health sciences).

9.2 Approach

The student team researched the policy landscape of firefighter turnout gear testing standards and the makeup of a moisture barrier and its required functions, and conducted health hazard and technical performance assessments on potential alternative materials. Although the challenge is multifaceted and could be evaluated from the manufacturing-side, consumer-side, or throughout the entire life cycle, the student team chose to specifically address the risk to wearers of firefighter turnout gear through a two-pronged policy and science approach.

9.2.1 Policy Approach

While considering alternatives to the PTFE laminate moisture barrier, the student team conducted a review of the NFPA 1971 performance standards used to evaluate firefighter turnout gear, in relation to the European Union (EN 469) and Australian (AUS) standards. The student team selected these standards as a comparison, guided by the IAFF, given their overall similarity to NFPA 1971. The NFPA is a self-funded non-government agency, led largely by industry stakeholders, with no official regulatory authority.[12] There are concerns that gear may be over-engineered due to over-prescriptive testing standards within NFPA 1971 that can be fulfilled by only PFAS materials currently. These potentially over-prescriptive standards go beyond what is required for safe and functional turnout gear, limiting functional PFAS-free alternatives.

Table 9.1 presents the comparison of the NFPA 1971 standards with the EN 469 and AUS standards. The student team used this table to assess differences in the standards.

As shown in Table 9.1, the NFPA 1971 standards include light degradation (also known as the "UV light test") and viral penetration tests, which are absent in the EN 469 and AUS standards. The moisture barrier sits inside the other two turnout gear layers and does not come into contact with ultraviolet (UV) light, making the UV light test unnecessary. Currently, a PTFE laminate is the only known material that meets this test, so removing the UV test would allow for PFAS-free moisture barrier options. In May 2021, the IAFF filed a Tentative Interim Amendment to remove the UV test from the

Table 9.1 Policy Standard Comparison for NFPA,[11] EN 469,[82] and AUS standards.[83]

Requirements	NFPA 1971[11]	EN 469 – level 2 (structural) European Union[82]	AUS (Australia)[83]
Heat/thermal resistance	Shall not shrink >10% in any direction.	At 5 minute exposure 180 °C: shall not shrink >5% in any direction. No ignition. No melting.	No standard
Water penetration (resistance)	Minimum water penetration resistance of 172 kPa (25 psi).	Minimum water resistance of 20 kPa	Using the rate of water pressure increase of 9.8 mbar min^{-1} and water temperature of 20 °C shall achieve ≥200 cm of H_2O
Tear resistance	Tear strength of not less than 22 N (5 lbf).	Tear strength of no less than 30 N	Coated materials shall give a tear strength equal to or greater than 25 N
Flame resistance	Char length <100 mm. Afterflame <2 seconds.	No standard	No standard
Cleaning shrinkage resistance	Cannot shrink more than 5%	Shall not shrink more than 3% (woven) Shall not shrink more than 5% (non-woven)	At a temperature of (2600 + 5) °C, no material shall melt, drip, ignite or shrink more than 5%
Liquid chemical resistance	No liquid can penetrate any sample for 1 hour	No penetration to innermost surface, index of repellency >80%	Clothing shall give >80% run off with no penetration to the innermost surface
Viral penetration resistance	Shall allow no penetration of the Phi-X-174 bacteriophage for at least 1 hour.	No standard	No standard
Light degradation resistance	Water shall not appear on the surface of the specimen when the garment is exposed to harsh light for 40 hours.	No standard	No standard

Table 9.1 (*Continued*)

Requirements	NFPA 1971[11]	EN 469 – level 2 (structural) European Union[82]	AUS (Australia)[83]
Water vapor resistance	No standard	Must be less than or equal to 30 m^2 Pa W^{-1}	After pre-treatment in accordance with clause 4.1.3, the complete component assembly shall achieve a water vapor resistance of less than 20.0 m^2 Pa W^{-1}

standards, but the NFPA denied it and subsequent appeals. In their appeal the IAFF stated, "The UV light degradation test for moisture barriers is illogical, not supported by science, and stands as a hurdle to advancing the state of the art in firefighter PPE."[13]

The viral penetration test is intended to test the fabric for resistance to liquid or bloodborne pathogens. The student team concluded that this test may also be unnecessary for the moisture barrier. Firefighters regularly wear their whole ensemble, *i.e.*, jacket, pants, and boots, whether they are responding to a fire or providing paramedical services. The DWR outer shell blocks viral particles from interacting with the moisture barrier, making the viral penetration test unnecessary.

The first draft of updated firefighter turnout gear performance standards, NFPA 1970, was released for public comment in late 2022. In the NFPA 1970 draft, the light degradation test is not included since firefighters have been advocating for its removal. However, there is no movement currently to reconsider the viral penetration test. Following public comment on the first draft, the NFPA anticipates that the second draft report will be posted for public review in the fall of 2023.[14] The student team submitted a public comment supporting the removal of the light degradation test, and recommending the additional removal of the viral penetration test.[15]

In conclusion, two NFPA 1971 tests – light degradation and viral resistance – are likely unnecessary for the moisture barrier and are prohibiting PFAS-free options from coming to the market. The remaining and relevant standards were used as technical performance requirements for the alternatives assessment, along with the addition of water vapor transmission, water vapor resistance, and pore size tests, which are not included in the NFPA 1971 standards, but are reasonable to ensure comfort and prevent steam burns.

9.2.2 Science Approach

The student team examined the components of the current expanded PTFE (ePTFE) moisture barrier to identify possible alternatives. One example of a moisture barrier currently used in firefighter turnout gear is the GORE®PARALLON™ Liner made by W. L. Gore and Associates. This example moisture barrier is composed of three layers (exterior layer, middle layer, and body-side layer) that each contain an ePTFE film laminate. The exterior layer contains NOMEX, a flame-resistant fabric containing a flame retardant coating, and is laminated with an ePTFE film; the middle layer is made of a KEVLAR/NOMEX blend and laminated with an ePTFE film; the body-side layer is made of a KEVLAR, NOMEX, LENZING fiber blend and laminated with an ePTFE film.[16] By breaking down this example moisture barrier into its three components, the student team grouped alternatives into three strategies: fibers, coatings, and laminates. It is possible that the fibers or laminates alone could replace the moisture barrier. While the primary purpose of the coating is to increase fire retardancy of the fibers, it is possible that some coatings could increase the hydrophobicity of certain fibers. Additionally, a combination of fibers, coatings, and laminates would produce an optimal moisture barrier. Coatings are polymeric layers applied to fibers. The current flame retardant coatings used in the moisture barrier are brominated. For this reason, a representative brominated flame retardant commonly used as a flame retardant coating, tetrabromobisphenol A (TBBPA), was evaluated.[17] Laminates are two independent layers of polymer and fabric held together by the adhesive properties of one of the layers or an adhesive. The current laminate in the moisture barrier is ePTFE.

9.2.2.1 *Technical Performance*

Properties of alternatives were compared to the following NFPA 1971 requirements: heat/thermal resistance, water penetration (resistance), liquid chemical resistance, and tear resistance. Direct comparisons of literature values to the NFPA standard values were made for tear resistance and water penetration resistance, but proxies were required for the other standards due to a lack of similar testing in the literature. For heat/thermal (resistance), the NFPA requirement is that the moisture barrier shall not shrink more than 10% in any direction. The proxy used for the heat/thermal (resistance) requirement was thermal resistance – the measure of the temperature difference by which a material resists heat flow. The NFPA liquid chemical resistance requirement is that no liquid tested (aqueous film-forming foam, 37% by weight sulfuric acid solution, fire-resistant hydraulic fluid, surrogate gasoline fuel, and chlorinated solution) should penetrate the moisture barrier for 1 hour. The proxy used for the liquid chemical resistance requirement was water contact angle – the measure of the angle a drop of liquid makes with a solid. A water contact angle greater than 90° indicates that the surface is hydrophobic and a water contact angle less than 90° means that the surface is hydrophilic.[18] A higher water contact angle will prevent the wetting of the material, indicating that the material is water resistant.

Breathability performance metrics including pore size, water vapor transmission, and water vapor resistance were evaluated for fibers and laminates, but not for coatings due to a lack of available data. One purpose of the moisture barrier is to prevent steam burns from sweat or water that gets trapped between the turnout gear and the skin. By optimizing the pore size so that water vapor can pass through the pores, but liquid water cannot, the breathability of the material can be increased and steam burns can be prevented. The water vapor transmission rate is a measure of the rate at which water vapor passes through a fiber or membrane at a certain humidity. Water vapor resistance is a requirement in the EN (European) protective clothing for firefighter standards. Water vapor resistance is the material's reluctance to allow water vapor to pass through.[19] The lower the water vapor resistance, the higher the breathability.

In addition to moisture barrier properties, the team assessed alternatives for their flammability or fire retardancy. If the moisture barrier is not flame/heat resistant, then it could decompose at the high temperatures firefighters endure. Due to a lack of data in the literature matching the NFPA standard tests for the flame retardant coatings, alternative performance metrics were evaluated for this component of the moisture barrier. The additional performance metrics were thermal decomposition (5%), defined as the temperature at which 5% by weight of the material decomposed, and char length.[20] The char length measured in the literature is different from the NFPA standard due to the difference in flame exposure time to the material.

Technical performance metrics are shown in Table 9.2.

9.2.2.2 Hazard Performance

Environmental and health hazards were evaluated for "bad actor chemicals" used in current moisture barriers (PTFE, PFASs associated with the manufacturing of PTFE, and TBBPA), and potential PFAS-free alternatives. In the assessments, hazards associated with group I human health endpoints like reproductive toxicity and carcinogenicity were prioritized because they are most relevant to the major risks faced by the firefighting population, like cancer.

Hazard rankings (very high, high, moderate, and low) were assigned for each chemical of interest for specific endpoints. A stepwise approach was used to assign rankings. First, hazard assignments were determined from the GreenScreen Guidance Document for Hazard Assessment (an open source hazard ranking system used by researchers, product formulators, and certifiers).[21] If no hazard assignments were determined using GreenScreen guidance, hazard assignments were extrapolated from European Chemical Agency (ECHA) dossiers on chemicals registered in the European Union (EU) through REACH (an EU regulation that requires hazard information on chemicals manufactured in or imported to the EU).[22] If there were still data

Table 9.2 Criteria used for ranking technical performance.

Metric	Optimal (high efficiency)	Sub-optimal (moderate efficiency)	Poor (low efficiency)
Liquid chemical resistance (water contact angle)	>100°	70–100°	0–69°
Pore size (diameter, μm)	>0.004 and <1	1–100	>100
Water vapor transmission (g per m^2 per day)	>1500	750–1500	<750
Thermal decomposition (5%) temperature (°C)	>300	150–300	<150
Char length	<2 mm	2 mm–1 cm	>1 cm

Metric	Meets standard (high efficiency)	Does not meet standard (low efficiency)
Water vapor resistance	>172 kPa (25 psi)	<kPa (25 psi)
Heat/thermal resistance	>0.008 $m^2 K W^{-1}$	<0.008 $m^2 K W^{-1}$
Tear resistance	>22 N	<22 N
Water vapor resistance (EU standard)	<30 $m^2 Pa W^{-1}$	>30 $m^2 Pa W^{-1}$

gaps, a literature search was conducted on the hazards associated with the chemical of interest.

Confidence rankings were also assigned to these hazard rankings. High confidences were assigned to hazard rankings clearly ranked in GreenScreen. Medium confidences were assigned to hazard rankings that were extrapolated from a range in GreenScreen (*e.g.*, given a range of very high–moderate hazard, the student team chose a high hazard), based on modeled or unverified values, or extrapolated from different GreenScreen ranking entities. Low confidences were assigned to hazard rankings determined by toxicity values (*e.g.*, no observed adverse effect levels/NOAELs or lethal dose/LD50s) provided by ECHA or a literature search which were then compared to known values from high hazard chemicals or ECHA designated low hazard rankings determined from preliminary animal studies. The low confidence interval stems from the fact that NOAELs were treated as LD50 values for comparison to GreenScreen, but NOAELs are not in fact LD50 values. Additionally, values were assigned low confidence if the values from ECHA or in the literature were questionable due to testing methodology, alignment with GreenScreen criteria, *etc.*

The results for the different chemicals were compared to determine whether the alternatives minimized hazards relative to the bad actors. The hazards of the potential alternatives were also assessed independent of the bad actor chemicals to determine whether the alternatives are viable on a scale of absolute risk.

While PTFE is considered inert with minimal hazards, the surfactants and monomers used to produce PTFE are toxic. Surfactants and unreacted monomers that are not properly removed after PTFE synthesis are of particular concern due to their presence throughout firefighter turnout gear.[23] Thus, in addition to PTFE, the student team also assessed hazard data for three PFASs used or formerly used in PTFE production and two additional well-known PFASs. These included tetrafluoroethylene (TFE) – the monomer used to produce PTFE; perfluorooctanoic acid (PFOA) – a legacy, "C8" PFAS formerly used as a surfactant in the production of PTFE; hexafluoropropylene oxide dimer acid (known as GenX) – a "C6" PFAS used as replacement for PFOA in PTFE production; perfluorooctanesulfonic acid (PFOS) – a legacy "C8" PFAS formerly used in firefighting foam; and perfluorodecanoic acid (PFDA) – a PFAS that has been detected in dust collected from firehouses. The student team also assessed hazard data for TBBPA.

The PFASs evaluated, TFE, PFOA, GenX, PFOS, and PFDA, and the brominated flame retardant, TBBPA, were all associated with at least one high or very high hazard in a group I Human Health Endpoint. More information on specific hazards associated with our "bad actor chemicals" can be found in Figure 9.2.

9.3 Strategies

9.3.1 Fibers

The base of each layer in the moisture barrier is the fiber. The current base layer is made from KEVLAR and NOMEX blends. KEVLAR and NOMEX, commonly referred to as aramid fibers, are synthetic fibers made from aromatic rings connected with amide linkages. These fibers tend to have a high tensile strength and heat resistance, which are important properties for firefighter turnout gear since it should be durable and resistant to high heat. Although these fibers are strong, their applicability as moisture barriers on their own is questionable. The student team chose to focus on three types of fibers – cellulose-based fibers, polyhydroxyalkanoates (PHAs), and graphene nanofibers. Cellulose-based fibers and PHAs were selected for their tunability and low toxicity, inferred from their use in a wide range of other clothing. Graphene nanofibers were selected for their porosity, tunability, and due to previously reported data that indicate their ability to produce fibers that are flame retardant.[24] If these fibers were lacking in the technical performance metrics, modifications could enhance their moisture barrier properties or their fire resistance.

9.3.1.1 Technical Performance

9.3.1.1.1 Cellulose-based Fibers. Cellulose is a polysaccharide consisting of a linear chain of D-glucose molecules. Cellulose is a primary component of plant cell walls and is secreted by some bacteria to form biofilms. Cellulose fibers, like viscose rayon and lyocell, are produced by dissolving natural materials like wood pulp and regenerating them *via* extrusion and precipitation.[25] Properties of cellulose fibers like their high

		Group I Human Endpoints			Group II and Group II* Endpoints			Ecotoxicity	Fate	Physical Hazard
		Carcinogenicity/ Mutagenicity	Develop. / Reproductive Toxicity	Endocrine Activity	Systemic Toxicity	Neurotoxicity	Skin, Eye, Respiratory Irritation/ Sensitization	Aquatic Toxicity Acute/ Chronic	Persistence/ Bioaccum.	Reactivity, flammability
PTFE + PTFE Monomer	PTFE	**M**[55]	DG	DG	DG	DG	DG	DG	**VH**[84]	DG
	TFE	**H**[84]	**VH**[85]	DG	**H**[86]	**H**[86]	**M**[86]	DG	**VH**[87]	**H**[88]
C8s/Legacy PFASs	PFOA	**H**[89]	**H**[89]	**H**[90]	**H**[91]	**H**[92]	**VH**[91,92]	**M**[92]	**VH**[93]	DG
	PFOS	**H**[94]	**H**[94]	**H**[95]	**H**[96]	DG	**M**[97]	**H**[97]	**VH**[98]	DG
Other PFASs	GenX	DG	**VH**[99]	DG	M[100]	DG	**VH**[100]	DG	**VH**[100]	**L**[100]
	PFDA	**M**[101]	**H**[101]	**H**[93]	**H**[102]	DG	**VH**[102]	DG	**VH**[103]	DG
Brominated FR	TBBPA	**H**[104]	**H**[105]	**H**[106]	DG	DG	DG	**VH**[105]	**VH**[107]	DG

Key

Hazard	High (H) / Very High (VH)	Moderate (M)	Low (L)	Data Gap (DG)
Confidence	**High**	Medium	*Low*	

Figure 9.2 Hazard assessment comparison for baseline substances (PTFE,[55,84] TFE,[84–88] PFOA,[89–93] PFOS,[94–98] GenX,[99,100] PFDA,[93,101–103] and TBBPA[104–107]).

breathability and biodegradability make them ideal greener candidates for the base layer of the moisture barrier in firefighter turnout gear; however, they are highly combustible. Since firefighter turnout gear is exposed to high heat from fires, a cellulose fiber will require modifications to be implemented as a moisture barrier in firefighter turnout gear. Critical moisture barrier properties like pore size can be achieved by modifying cellulose to make cellulose nanofibers.[26] The following paragraphs discuss different cellulose-based fibers that were evaluated based on the technical performance criteria mentioned previously.

Cotton had a low water contact angle, but had a high water vapor transmission rate and low water vapor resistance. Additionally, cotton met the tear resistance requirements. Although cotton is a breathable fabric that is comfortable for firefighters to wear, modification of cotton is necessary to increase the water contact angle and the thermal resistance of the material.

Cellulose nanofibers had many data gaps in the thermal resistance, water penetration resistance, and tear resistance standards. Although the pore size was optimal, the material's water contact angle below 90° makes it hydrophilic and a nonideal moisture barrier.

Lyocell, also marketed as Tencel, is a semisynthetic fabric made from regenerated cellulose. It has similar properties to cotton and linen; however, it is reported to have greater breathability and be 50% more absorbent than cotton. It is a "greener" fabric compared to cotton and other cellulose-based fabrics due to safer chemicals used in the manufacturing process.[27] Lyocell performed well in the breathability performance metrics including pore size, water vapor transmission rate, and water vapor resistance; however, there were data gaps for the thermal resistance and water contact angle. Before Lyocell could be used in a moisture barrier, measures should be taken to decrease the flammability of the fiber and increase the liquid chemical resistance.

Viscose rayon is a semisynthetic fiber made from cellulose. It is semisynthetic due to the chemicals like sodium hydroxide and carbon disulfide which are involved in the manufacturing process. Similar to other cellulose fibers, viscose rayon is absorbent, breathable, and lightweight. It performed well in the thermal resistance test; however, this divergence from other cellulose fibers may be explained by the difference in humidity percentage employed for the thermal resistance measurements.

Overall, the cellulose fabrics failed to meet a number of technical performance metrics, ranging from a low water contact angle to low tear resistance. One way to increase the water contact angle is to treat the fabric with a hydrophobic compound. For example, incorporating betulin and betulinic acid into cellulose-based fibers resulted in an increase in the water contact angle to >90°.[28] Improvements in the hydrophobicity, strength, and breathability of the fibers could be achieved by blending with synthetic fibers, which are generally more durable and hydrophobic, with the cellulose fibers. For example, the breaking strength of cotton fibers was improved by increasing the polyester: cotton fiber blend ratio.[29]

9.3.1.1.2 Short Chain PHAs. Polyhydroxyalkanoates (PHAs) are a family of biodegradable polyesters. They are produced by microorganisms, like bacteria, for carbon and energy storage purposes.[30] Their structural variability leads to PHAs with a variety of characteristics ranging from crystalline, stiff substrates to elastic material.[31] Although unmodified PHAs could be viable as moisture barriers due to their hydrophobicity, one detriment of these materials is their low thermal decomposition temperature. Thus, modification of PHA fibers is required to decrease their flammability.

Polyhydroxybutyrate (PHB) is a representative PHA the student team chose to investigate since it is well characterized in blended fibers and films.[32] PHB performed well in the liquid chemical resistance test since it has a high water contact angle; however, there were data gaps for water penetration resistance and thermal resistance.

9.3.1.1.3 Graphene Nanofibers. Graphene is an allotrope of carbon consisting of repeated six-membered carbon rings. This two-dimensional material's utility stems from its lightweight structure and its tunability *via* doping with various elements and functional groups. Investigations are being conducted into graphene's applicability as a semiconductor in computer chips, in biomedical sensors, in battery storage for wind and solar power, and most importantly, in fibers.[33] Graphene can be spun into fibers by electrospinning, a process where a liquid droplet is electrified to generate a jet that is then stretched.[34]

The technical performance of graphene nanofibers was average compared to the other fibers. Although the material is hydrophobic with a water contact angle above 90°, there were data gaps for thermal resistance and tear resistance. In addition, the thermal resistance and water vapor transmission of graphene nanofibers were not as high as those of other fibers. However, the pore size of the material was optimal and other properties could be improved with slight modifications. Other areas of concern for graphene are scalability, cost, and conductivity.

Currently, graphene nanofibers and other graphene materials are synthesized on a much smaller scale than the material required for a moisture barrier. Commercially available graphene is synthesized through a time-consuming and costly process called chemical vapor deposition (CVD), in which gasses react and deposit on a surface to make a graphene layer.[35] Graphene can cost \$67 000–\$200 000 per metric ton, although this price range could be for lower quality graphene.[31] It is important to note this cost is for graphene before nanospinning, so graphene nanofibers could be more expensive. Research into a faster and cheaper graphene synthesis method is under investigation by researchers at Rice University. The researchers make graphene by flash Joule heating, which heats carbon-based material at high temperature for a few seconds, resulting in the breaking of the chemical bonds.[36] After heating, gasses escape and the carbon reassembles into graphene. Although more research is needed to determine the scalability of this process, it could be a viable alternative for the mass production of graphene nanofibers.

Another area of concern for graphene is conductivity. Graphene is conductive due to the repeating carbon double bonds in its structure. Although conductivity is an asset when graphene is used in biosensors and other electronic devices, it could be a detriment in firefighter turnout gear.[33] Should firefighters come into contact with loose wires or electrified materials, a conductive fiber could lead to electrical burns. Although the moisture barrier is the middle layer of the fabric and does not come into contact with the skin, this trait should be considered moving forward to ensure that it does not present an additional hazard.

Figures 9.3 and 9.4 show performance data and ranking for the current moisture barrier fabrics and potential alternatives.

9.3.1.2 Hazard Assessment

9.3.1.2.1 Cellulose-based Fibers. A hazard assessment was conducted on the cellulose monomer (the base of all cellulose-based fibers) to serve as a proxy for the hazards associated with cellulose-based fibers. The cellulose monomer is a known sensitizer-induced asthmagen in occupational settings.[37] This hazard is relevant in the manufacturing of cellulose-based fibers; however, the risk related to asthma is essentially null once the

	Compound	Water Penetration Resistance	Liquid Chemical Resistance Water Contact Angle (degree)	Thermal Resistance (m^2*K/W)	Tear Resistance (Newtons)
Current Fabrics	KEVLAR (single layer)		61.2-66.2[108, 109]	0.008[110]	3620 N/(mm²)*[111]
	KEVLAR/Wool Blend (single layer)			0.011[109]	
	NOMEX		144.7, 117.4[112]		113[113]
Graphene	Graphene Nanofibers		95-100[114]	0.09 m K/W*[115]	
Cellulose Based Fabrics	Cellulose Nanofibers		85.9[116]		22.78 N/(mm²)*[117]
	Lyocell (Tencel)				21.16[118]
	Cotton		40.2-43.9[119]	0.171[120]	10.79[121]
	Viscose rayon			0.189[120]	308 N/(mm²)*[122]
Biobased Fabric	PHA/PHB fibers		118[123]		40 N/(mm²)*[124]

Key	Efficiency	High	Moderate	Low

Figure 9.3 Technical performance summary for three baseline fibers (KEVLAR,[108–111] KEVLAR/wool,[109] and NOMEX[112,113]) and proposed alternative fibers (graphene nanofibers,[114,115] cellulose nanofibers,[116,117] lyocell,[118] cotton,[119–121] viscose Rayon,[120,122] PHA/PHB fibers[123,124]).

	Compound	Pore Size (diameter, μm)	Water Vapor Transmission Rate (g/(m²day))	Water Vapor Resistance (m²Pa/W)
Existing Fabric	*KEVLAR (single layer)*	9.0-12.8[125]	863.4[126]	9.2[109]
	KEVLAR/Wool Blend (single layer)			7.3[109]
	NOMEX	10-100[127]	7456[128]	3.48[129]
Graphene	Graphene Nanofibers	0.003-0.050[130,131]	848.22[112]	
Cellulose Based Fibers	Cellulose Nanofibers	0.02-0.3[132]	20-160[133]	
	Lyocell (Tencel)	0.001-0.1[134]	~1000[135]	3.35[136]
	Cotton		1441.42[137]	1.85[138]

Key	Efficiency	High	Moderate	Low

Figure 9.4 Additional technical performance for three baseline fibers (KEVLAR,[109,125,126] KEVLAR/wool blend,[109] and NOMEX[127–129]) and proposed alternative fibers (graphene nanofibers,[112,130,131] cellulose nanofibers,[132,133] lyocell,[134–136] and cotton[137,138]).

monomer is synthesized into a fiber. Thus, the risk related to asthma is irrelevant to the firefighter population.

As cellulose is a common building material of plant cell walls, people frequently consume it in vegetables, so data gaps in the hazard assessment for cellulose were determined to not be of high concern. Thus, cellulose was found to be a low risk to firefighters.

9.3.1.2.2 Short Chain PHAs. A hazard assessment was conducted on polyhydroxybutyrate. One animal study found no association between PHB and cancer in rats, indicating a low hazard with low confidence.[38] Preliminary research found a low modeled half-life of PHB in fish of 0.504 days, suggesting low hazard with low confidence for aquatic toxicity.[39] Based on the predicted half-life of 4.47 days in soil, PHB was designated as low hazard with high confidence for environmental persistence.[39] Based on its $\log K_{ow}$ of 0.31, PHB was designated low hazard with high confidence for bioaccumulation.[39]

Even though the use of short chain PHAs in fiber production is novel, they are biopolymers produced by microorganisms and thus are projected to have few environmental and health hazards. Although there are considerable data gaps associated with short chain PHAs, these gaps were found to be of limited concern.

9.3.1.2.3 Graphene. Due to limited data on graphene nanofibers, a hazard assessment was conducted on graphene, specifically graphene nanoplatelets. Graphene is "not rapidly biodegradable" and considered "harmful to aquatic life with long lasting effects".[40] However, preliminary animal studies found no skin sensitization or irritation and no eye irritation associated with graphene nanoplatelets.[40] Additional animal studies also found no genetic toxicity.[40] An animal study investigating the impact of repeat doses of graphene nanoplatelets through inhalation found no observed impact on rats.[40]

Overall, preliminary animal studies suggest low toxicity with low confidence related to group I and II endpoints, indicating that from a hazard standpoint graphene is a viable alternative. There are concerns of persistence and aquatic toxicity, but these concerns are of less relevance to firefighters once graphene is incorporated into turnout gear. Figure 9.5 shows hazard data and ranking for the potential alternative fibers.

9.3.2 Coatings

Fibers alone are not sufficient to act as a moisture barrier in firefighter turnout gear due to their low thermal resistance and/or small water contact angle. Fire retardant coatings can combat this by increasing the thermal resistance of the fibers, thereby decreasing their flammability. It is also possible that fire retardant coatings could increase the hydrophobicity of the fiber, improving its moisture barrier properties. For example, cellulose-based fibers have many free hydroxyl groups, which are reactive and hydrophilic and lead to a low water contact angle. Flame retardant coatings react with the hydroxyl groups, essentially "blocking" these sites, resulting in hydrophobic groups pointing away from the fiber. An example of this is shown below where the flame retardant Exolit 5060 PK reacts with a cellulose-based fiber (see Figure 9.6).

	Group I Human Endpoints			Group II and Group II* Endpoints			Ecotoxicity	Fate	Physical Hazard
	Carcinogenicity/ Mutagenicity	Develop. / Reproductive Toxicity	Endocrine Activity	Systemic Toxicity	Neurotoxicity	Skin, Eye, Respiratory Irritation/ Sensitization	Aquatic Toxicity Acute/ Chronic	Persistence/ Bioaccum.	Reactivity, flammability
Cellulose Monomer	DG	DG	DG	DG	DG	H[37]	DG	DG	DG
PHB	*L*[139]	DG	DG	DG	DG	DG	*L*[39]	L[39]	DG
Graphene nanoplatelets	*L*[40]	DG	DG	*L*[40]	DG	*L*[40]	M[40]	*H*[40]	*L*[40]

Key	Hazard	High (H) / Very High (VH)	Moderate (M)	Low (L)	Data Gap (DG)
	Confidence	**High**	Medium	*Low*	

Figure 9.5 Hazard assessment for proposed alternative fibers (cellulose,[37] short chain PHAs,[39,139] and graphene nanoplatelets[40]).

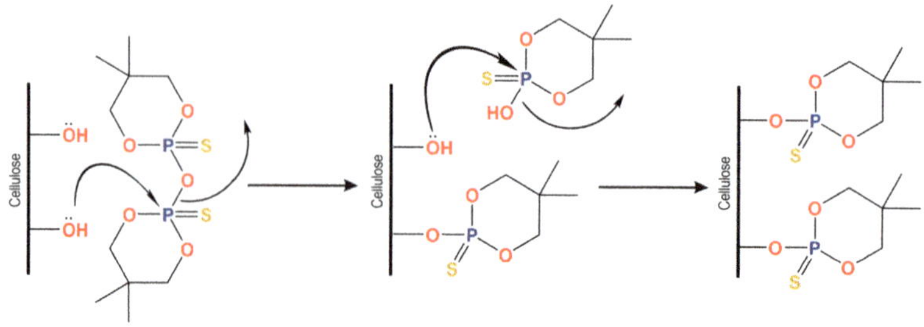

Figure 9.6 Schematic showing the reaction of Exolit 5060PK with hydroxyl groups on a cellulose fiber.[47]

9.3.2.1 Technical Performance

9.3.2.1.1 Silicon-based Flame Retardants. Silicon/silicone materials have been produced commercially for over 80 years and are widely used in polymer products, construction, textiles, and electronics. One of the most common silicon materials is polydimethylsiloxane (PDMS), a polymer that is incorporated into electrical wires and cables. The stability and flame retardancy of these compounds is related to the strength of the Si–O bonds, which are inherently strong, and the flexibility of $-[Si-O]_x-$ segments.[41] Silicon-based flame retardants are incorporated into textiles and building materials to decrease flammability. The student team investigated three silicon-based flame retardants as coatings that could enhance the moisture barrier properties of fibers: two chemicals referred to as BD and BDD, and tetraethoxysilane (TEOS)/polydimethylsiloxane (HPDMS).

The novel silicon-based flame retardants BD and BDD (see Figure 9.7 for structures) are examples of Schiff bases, which are commonly used in drug discovery and thermo/photochromic materials. Schiff base applications as fire retardant coatings are investigated due to their cross-linking properties and ability to form a char layer.[42] In one study, BDD decomposed at a higher temperature than BD, indicating that BDD can withstand higher temperatures, which is crucial due to the high temperatures firefighter turnout gear must endure.[43] Although more investigation needs to be conducted into the water contact angle to determine whether this fiber-coating combination could act as a moisture barrier; the hydrophobic groups of these compounds, like benzene rings, suggest that this fire retardant/coating would be hydrophobic.

TEOS/HPDMS on cotton fibers performed the best of all the flame retardant coatings investigated. The water contact angle was over 160° and the thermal decomposition temperature was the highest at 264 °C. Although this temperature is lower than the temperature firefighters experience when combating structural fires (300 °C), it is possible that a combination of this fire retardant with another fiber (cellulose nanofibers, lyocell, *etc.*) would increase the thermal decomposition temperature.[20] Although the char length was high, it was lower than the other fiber-coating combinations (see Figure 9.8).

BD

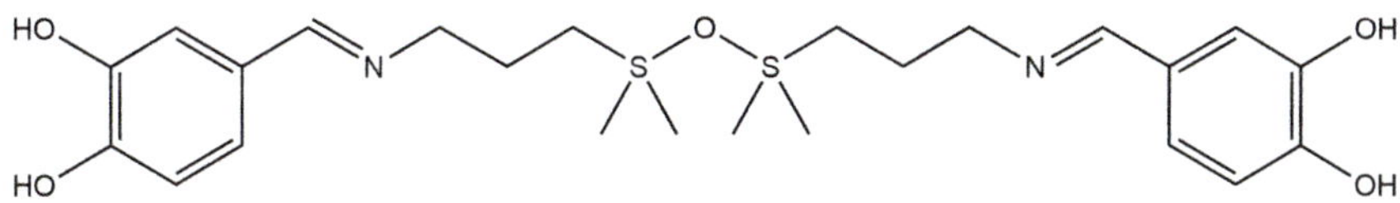

BDD

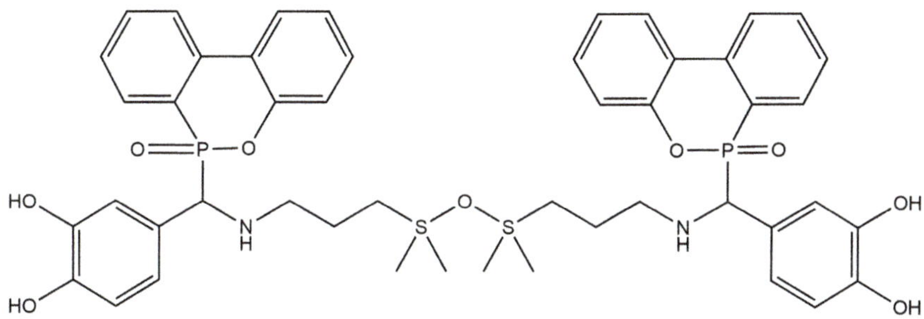

Figure 9.7 Structure of two novel silicon-based flame retardants, BD and BDD. Reproduced from ref. 43 with permission from American Chemical Society, Copyright 2020.

	Compound	Liquid Chemical Resistance Water Contact Angle (°)	Thermal Decomposition (5%) Temperature (°C)	Char Length (cm)*
Phosphorus Based Flame Retardants	Viscose Rayon + Exolit 5060 PK	Stable in Acidic and Basic Solution[47]		
	Cotton + DOPO		220[46]	12.2[46*]
Silicon Based Flame Retardants	Cotton + BD		162[43]	
	Cotton + BDD		225[43]	
	Cotton + TEOS/HPDMS	>160[20]	264[20]	8.5[20*]

Key	Efficiency	High	Moderate	Low

Figure 9.8 Technical performance of five proposed fiber/coating alternatives (viscose rayon/Exolit 5060 PK,[47] cotton/DOPO,[46] cotton/BD,[43] cotton/BDD,[43] cotton/TEOS/HPDMS[20]).

9.3.2.1.2 Phosphorus-based Flame Retardants. Phosphorus-based flame retardants are currently incorporated into a variety of materials like polyethylene terephthalate (PET) fibers, polyurethane foams, furniture material, and epoxy resins.[44] In Europe, they are being used to replace halogenated flame retardants like brominated flame retardants and polychlorinated biphenyls.[45] Similarly, they could replace the brominated flame retardants in the moisture barrier. A literature analysis revealed two phosphorus-based flame retardants that, in addition to decreasing the flammability of the fiber, could act as moisture barriers.

9,10-Dihydro-9-oxa-10-phosphaphenanthrene-10-oxide (DOPO) is a phosphorus-containing compound incorporated into epoxy resins to decrease flammability, but there are few reports of it being incorporated into fibers. One report indicates that modification of DOPO with various silicon compounds can increase fire retardancy and allow for more efficient incorporation into fibers. One example of this is 9,10-dihydro-9-oxa-10-phosphaphenanthrene-10-oxide-vinyltrimethoxysilane (DOPO-VTS), which was incorporated into cotton fibers and showed a flame-retardant effect. Another example is DOPO–PiP–Si, which had a thermal decomposition temperature of 220 °C when coated onto cotton fibers.[46] Although this temperature is below 300 °C, it is possible that a different fiber or different modification to DOPO would result in a larger decrease to fiber flammability. In addition, the water contact angle should be assessed to determine whether this fiber-coating combination could act as a moisture barrier since its chemical structure suggests that this fire retardant/coating would be hydrophobic.

Exolit 5060 PK is a phosphorus-based flame retardant that is incorporated into viscose rayon by the chemical company Clariant.[47] Our review identified little performance data for this flame retardant. It was reported to be stable in acidic and basic media;[47] this non-reactivity is an ideal characteristic for a moisture barrier. In addition, the fire retardant complies with fire safety standard DIN-4102-1, a German fire standard specific to buildings. More quantitative testing should be performed to determine the effectiveness of this fiber-coating as a moisture barrier. Figure 9.8 shows the comparative technical performance assessment for both phosphorus-based and silicon-based flame retardants.

9.3.2.2 *Hazard Assessment*

9.3.2.2.1 Silicon-based Flame Retardants. Due to the lack of information for the novel silicon-based flame retardants BD and BDD, the chemicals 3,4-dihydroxybenzaldehyde, 9,10-Dihydro-9-oxa-10-phosphaphenanthrene-10-oxide, and 1,3-bis(3-aminopropyl)-1,1,3,3-tetramethyldisiloxane; which are used in the synthesis of BD and BDD, were assessed. 3,4-Dihydroxybenzaldehyde and 1,3-bis(3-aminopropyl)-1,1,3,3-tetramethyldisiloxane are associated with severe skin and eye damage, and 1,3-bis(3-aminopropyl)-1,1,3,3-tetramethyldisiloxane was found to be very toxic to aquatic life.[48–50] 9,10-Dihydro-9-oxa-10-phosphaphenanthrene-10-oxide was not associated with genotoxicity, skin and eye irritation, or toxicity to any organs; however, it was found to be a potential skin sensitizer.[48]

TEOS was found to cause skin and eye irritation, and cause damage to the respiratory tract through inhalation.[51] HPDMS was found to cause skin and eye irritation.[52]

Given the data gaps for these chemicals, proceeding with caution is advised when implementing silicon-based flame retardants into moisture barriers for turnout gear. Even though further research into hazards of these silicon-based chemicals is recommended, current data suggest that they could be viable alternatives to brominated flame retardant coatings in the moisture barrier. Additionally, high hazards were only related to skin and eye irritation (see Figure 9.9), which are less of a concern to firefighters once these reagents are synthesized into their respective silicon fire retardant.

9.3.2.2.2 **Phosphorus-based Flame Retardants.** To assess the hazards of phosphorus-based flame retardants, the project team considered hazard data for novel FRs Exolit 5060 PK and DOPO, as well as data for diazinon and chlorpyrifos, two well-studied phosphorus-based pesticides. The pesticides were included in order to provide information on the "worst case hazard scenario" for phosphorus-based chemicals and to fill in data gaps for novel phosphorus fire retardants.

Preliminary animal studies on Exolit 5060 PK found no risk of mortality and no toxic effects related to reproductivity at the highest doses administered, indicating low hazard associated with reproductive and systemic toxicity.[53] Experimental studies did not find an association between Exolit 5060 PK and skin irritation or sensitization.[54] The bioconcentration factor for Exolit 5060 PK indicates that it does not bioaccumulate; however, Exolit 5060 PK may cause long lasting harmful effects to aquatic life.[54] DOPO also presents potential hazards, as it is toxic to aquatic life with long lasting effects and may cause an allergic skin reaction.[48,49]

As Exolit 5060 PK and DOPO are understudied chemicals, the student team used well-studied chemicals to inform data gaps (see Figure 9.9). Diazinon is a probable carcinogen and potential endocrine disruptor, found to be neurotoxic, and is persistent in the environment.[55–63] Chlorpyrifos is a potential endocrine disruptor, causes damage to organs, and is known to be developmentally toxic.[64–66] Given these concerning hazards, further research on novel phosphorus-based flame retardants like Exolit 5060 PK and DOPO is suggested before applying them as alternatives.

9.3.3 Laminates

Laminates in the moisture barrier are expanded polymers, which are made by incorporating a blowing agent during the manufacturing process. This blowing agent creates a cellular structure[67] with increased porosity that allows for airflow through the material but prevents liquid water from penetrating the membrane.[68] Similar to PTFE, other polymers like polyethylene

		Group I Human Endpoints			Group II and Group II* Endpoints			Ecotoxicity	Fate	Physical Hazard
		Carcinogen. / Mutagenicity	Develop. / Reproductive Toxicity	Endocrine Activity	Systemic Toxicity	Neurotoxicity	Skin, Eye, Respiratory Irritation/ Sensitization	Aquatic Toxicity Acute/ Chronic	Persistence/ Bioaccum.	Reactivity, flammability
Silicon Flame Retardants	3,4-Dihydroxybenzaldehyde*	DG	DG	DG	DG	DG	**H**[140]	DG	DG	DG
	9,10-dihydro-9-oxa-10-phosphaphenanthrene-10-oxide*	*L*[48]	DG	DG	**L**[48]	DG	*M*[48]	*L*[48]	*L*[48]	DG
	1,3-Bis(3-aminopropyl)-1,1,3,3-tetramethyldisiloxane*	DG	DG	DG	DG	DG	VH[50]	VH[50, 141]	DG	DG
	TEOS**	DG	*L*[86]	DG	**VH**[51]	DG	**H**[51]	DG	*L*[142]	**M**[143]
	HPDMS***	DG	DG	DG	DG	DG	**H**[144]	DG	*L*[52]	DG
Phosphorus Based Flame Retardants	Exolit 5060 PK	DG	*L*[54]	DG	*L*[54]	*H*[145]	*L*[53]	**M**[53]	*M*[145,54]	**L**[53]
	DOPO	DG	DG	DG	DG	DG	**H**[49]	**H**[48]	DG	DG
Phosphorus Based Pesticides	Chlorpyrifos	DG	**H**[65]	**H**[66]	**VH**[63]	H[93]	**H**[146]	**VH**[147]	**VH**[64]	DG
	Diazinon	**H**[59]	**M**[56]	**H**[63]	**H**[60]	H[58]	**H**[61]	**VH**[57,55]	**VH**[62]	DG

*Three chemicals used in BD/BDD synthesis

Key

Hazard	High (H) / Very High (VH)	Moderate (M)	Low (L)	Data Gap (DG)
Confidence	**High**	Medium	*Low*	

Figure 9.9 Hazard assessment for silicon flame retardant reactants (3-4-dihydroxybenzaldehyde,[140] 9-10-dihydro-9-oxa-10-phosphaphenanthrene-10-oxide,[48] 1,3-bis(3-aminopropyl)-1,1,3,3-tetramethyldisiloxane,[50,141] TEOS,[51,86,142,143] HPDMS[52,144]) and two phosphorus flame retardants (Exolit 5060 PK,[53,54,145] DOPO,[48,49] Chlorpyrifos,[63–66,93,146,147] and Diazinon[55–63]).

and polyurethane can be expanded, allowing them to be applied as a laminate and employed as a moisture barrier. Gore-Tex is currently looking into expanded polyethylene (ePE) membranes as an alternative to the ePTFE moisture barrier.[69]

9.3.3.1 Technical Performance

9.3.3.1.1 Polyethylene. Polyethylene is a polymer composed of a repeating network of carbons and hydrogens with a similar structure to PTFE. It is commonly used in chemical-resistant flooring, packaging, and containers for a variety of products. Depending on the synthesis method, high and low density polyethylene can be made.[70] Low density polyethylene is of particular interest since short and long branches in the polymer chain prevent closed packing and lead to the formation of a flexible material. These properties, as well as polyethene's low thermal conductivity, make it a potential alternative to the PTFE moisture barrier.

Polyethylene performed well for two of the standards examined: the water contact angle was 126° and the tear resistance was above the NFPA requirement. However, polyethylene ranked poorly for the pore size and water vapor transmission rate tests. Additionally, there were data gaps for the thermal resistance and water penetration resistance of the material.

9.3.3.1.2 Polyurethane. Polyurethane is a polymer that was first developed in the 1930s. It has properties similar to rubber.[71] Over time, research into polyurethane revealed that it can be modified to make foams and fine threads. It is often found as insulation in houses, in floor coatings, and in tires. Due to its hydrophobicity, durability, and its use as a DWR in outdoor wear, it could be an ideal alternative to the ePTFE membrane.

Polyurethane has a high water contact angle of 127° and a similar thermal resistance to ePTFE, but the breathability of polyurethane varies, due to variability in pore size and water vapor transmission rate.[72] In addition, there were data gaps in the water vapor resistance and the tear resistance of the material. Although a value was found for the water penetration resistance, the team was unable to compare the value due to a difference in the test reported in the literature.

Figures 9.10 and 9.11 show performance data and ranking for the current ePTFE laminate and potential alternatives.

9.3.3.2 Hazard Assessment

9.3.3.2.1 Polyethylene. A hazard assessment was conducted on the polyethylene (PE) polymer and the PE monomer ethylene. PE is a suspected asthmagen and is known to be persistent in the environment.[73,74] The monomer ethylene demonstrates evidence of carcinogenic effects but these effects were not sufficient enough to classify it as a carcinogen.[75] Ethylene was found to be neurotoxic and may cause respiratory tract irritation

Compound	Water Penetration Resistance (kPa)	Liquid Chemical Resistance Water Contact Angle (°)	Thermal Resistance (m^2*C/W)	Tear Resistance (Newtons)
ePTFE		125.5[148]	0.0671[149]	
PU	3.58-11.1[150]	70[151] 80-120[152]	0.0659[148]	
ePE		126[153]		39.29[154*]

Key: Efficiency | High | Moderate | Low

Figure 9.10 Technical performance of one baseline laminate (ePTFE[148,149]) and two proposed alternative laminates (PU[148,150–152] and ePE[153,154]). ePTFE = expanded polytetrafluoroethylene, PU = polyurethane, ePE = expanded polyethylene.

Compound	Pore Size (diameter, μm)	Water Vapor Transmission Rate (g/(m^{2}day))	Water Vapor Resistance (m^{2}Pa/W)
ePTFE	0.02-10[155,156]	5000-8000[155]	
PU	677[157]	~1100-4500[159]	
PU	50[158]	790-1050[72]	
ePE	5-250[160]	1.55-7.13[161] 8315[162]	

Key: Efficiency | High | Moderate | Low

Figure 9.11 Additional technical performance of one baseline laminate (ePTFE[155,156]) and two proposed alternative laminates (PU[72,157–159] and ePE[160–162]). ePTFE = expanded polytetrafluoroethylene, PU = polyurethane, ePE = expanded polyethylene.

through inhalation.[63,76] Additionally, ethylene was found to be harmful to aquatic life with long lasting effects.[77]

While there are potential concerning health impacts related to ethylene, once ethylene is synthesized into a polymer it poses a significantly lower risk to the firefighter population. As there are no group I human endpoints associated with the polymer, and it is not likely to degrade into its monomer form, PE is likely to be a viable laminate alternative to the PTFE moisture barrier. Since PE is one of the most commonly used plastics in the world and widely studied, data gaps are suspected to be of limited concern rather than gaps in knowledge.

9.3.3.2.2 **Polyurethane.** A hazard assessment was conducted on the polyurethane (PU) polymer and two reagents used to synthesize PU – glycerol and methylene diphenyl diisocyanate (see Figure 9.12). According to air quality standards by the Department of Defense, PU concentrations in the air were recommended to be below 0.25 $mg L^{-1}$.[78] This concentration was assigned to be a high hazard with low confidence for acute systemic toxicity. This hazard level was extrapolated from the GreenScreen designated high hazard range for median lethal doses with the same units. The PU reagents were found to be skin and respiratory sensitizers and impact the reproductive system.[79,80] One of the reagents, methylene diphenyl diisocyanate, is suspected to be cancer-causing.[80]

Like PE, there are potential concerning health impacts related to the reagents used to synthesize PU. However, once these reagents are synthesized into their polymer form, they pose a significantly lower risk to the firefighter population. The polymer form does not pose risks to any group I endpoints and is not likely to readily degrade into its reactants. As PU is one of the more commonly used materials in household furnishing and has been researched for decades, hazard endpoints with data gaps are likely because studies have not indicated hazards rather than information gaps.

9.4 Recommendations/Conclusions

The alternatives were evaluated based on technical performance metrics identified through the NFPA 1971 Standards critique. Fibers, flame retardant coatings, and laminates were each evaluated individually. Through the technical assessment, it was determined that the fibers alone are not sufficient to be used as a moisture barrier, due to low water contact angles or low thermal resistance. However, the addition of a flame retardant coating could potentially increase the water contact angle and thermal resistance based on technical assessments of the silicon and phosphorus-based flame retardants, with TEOS/HPDMS performing the best on the tests in the NFPA 1971 standards. However, there were multiple data gaps in the technical performance assessment, and more information is needed to properly assess the alternatives as moisture barriers, specifically related to water penetration resistance. A hydrostatic pressure test could be used to further evaluate water penetration resistance. For laminates, the technical assessment showed that polyethylene met more NFPA 1971 standards than polyurethane, but more research is needed to further evaluate thermal resistance.

Alternatives were also evaluated *via* a stepwise hazard assessment using a hazard assessment guidance document, chemical dossiers, and the literature. Based on the existing literature, all three alternative fibers appear to

	Group I Human Endpoints			Group II and Group II* Endpoints			Ecotoxicity	Fate	Physical Hazard
	Carcinogen. / Mutagenicity	Develop./ Repro. Toxicity	Endocrine Activity	Systemic Toxicity	Neurotoxicity	Skin, Eye, Respiratory Irritation/ Sensitization	Aquatic Toxicity Acute/ Chronic	Persistence/ Bioaccumu.	Reactivity, flammability
Polyethylene	*L*[163]	DG	DG	DG	DG	*M*[73]	DG	**VH**[74]	DG
Polyethylene Monomer	**M**[75]	*L*[76]	DG	**M**[76]	*H*[63]	DG	**M**[166]	DG	**H**[77]
Polyurethane	*L*[164]	DG	DG	*H*[78, 165]	DG	DG	DG	DG	DG
Methylene Diphenyl Diisocyanate	**M**[80]	**M**[80]	DG	**H**[80]	DG	**H**[80]	DG	DG	DG
Glycerol	*L*[79]	**M**[79]	DG	M[79]	DG	H[79]	*L*[79]	DG	DG

Key

Hazard	High (H) / Very High (VH)	Moderate (M)	Low (L)	Data Gap (DG)
Confidence	**High**	Medium	*Low*	

Figure 9.12 Hazard assessment for proposed alternative laminates (polyethylene[73,74,163] and polyurethane[78,164,165]) and chemicals involved in alternative laminate synthesis (polyethylene monomer,[63,75–77,166] methylene diphenyl diisocyanate,[80] and glycerol[79]).

present low hazards and data gaps likely present little concern. Cellulose fibers may present a manufacturing risk and graphene is potentially persistent in the environment; however, all of the fibers are viable safer alternatives. Regarding coatings – silicon-based flame retardants present a potential manufacturer risk (skin and eye irritation) and include data gaps associated with hazards for group I endpoints. Phosphorus-based flame retardants are associated with multiple high hazards for group I endpoints like carcinogenicity and endocrine disruption. Due to the comparative hazards, the student team recommends silicon-based flame retardants, although data gaps suggest further research on hazards before implementation. For laminates, both reagents used to synthesize both polyethylene and polyurethane present moderate hazards relating to carcinogenicity and a potential manufacturer risk. Due to extensive past research on these polymers, associated data gaps likely represent no evidence of hazards.

Considering findings from both the technical performance and hazard assessments together, the student team recommended the following. Regarding the science-side of the solution, the silicon-based flame retardant coatings could be incorporated with any of the three fibers: cellulose fibers, short-chain PHAs, or graphene. Alternatively, the polyethylene laminate could be applied to one of the evaluated fibers. An ideal moisture barrier would combine all three components – fibers, coatings, and laminates – however, due to a lack of technical performance data for all three components combined, the student team could not draw conclusions for a moisture barrier composed of a fiber, coating, and laminate. Concerning the policy-side, the student team advises the NFPA to remove the light degradation and viral penetration tests and keep the remaining standards. These tests are not required in the European Union nor the Australian standards, which implies that these tests are non-essential to the performance of the gear. The light degradation and viral penetration tests result in an overengineered moisture barrier and limit the viability of PFAS-free options.

9.5 Importance, Innovation, and Impact

Concern for firefighter safety from toxic chemicals is an increasingly highlighted issue. For example, at the 2023 IAFF Affiliate Leadership Training Summit, workshops aimed at informing participants about the dangers of PFASs were held and BURNED, a documentary about the deadly connection between PFASs and cancer, premiered. Additionally, at the 2023 IAFF Legislative Conference, President Joseph Biden spoke to firefighters about his commitment to addressing PFAS toxicity in firefighter turnout gear.

The students approached this challenge by recommending substitutions to each of the moisture barrier components (fibers, coatings, and laminates); however, it may be necessary to completely redesign the turnout gear entirely. Further evaluation of the NFPA and other standards could be a viable next step and may reveal overengineered standards that prevent viable alternatives from being considered.

The IAFF plans to continue working on this issue, and recently secured a grant to research the next generation of turnout gear. As previously mentioned, there is a current policy push to remove the light degradation test from the NFPA 1970 standards. The IAFF is working to develop safer alternatives without waiting for the NFPA to move the needle.

References

1. G. F. Peaslee, J. T. Wilkinson, S. R. McGuinness, M. Tighe, N. Caterisano, S. Lee, A. Gonzales, M. Roddy, S. Mills and K. Mitchell, *Environ. Sci. Technol. Lett.*, 2020, **7**, 594–599.
2. P. A. Demers, D. M. DeMarini, K. W. Fent, D. C. Glass, J. Hansen, O. Adetona, M. H. G. Andersen, L. E. B. Freeman, A. Caban-Martinez, R. D. Daniels, T. R. Driscoll, J. M. Goodrich, J. M. Graber, T. L. Kirkham, K. Kjaerheim, D. Kriebel, A. S. Long, L. C. Main, M. Oliveira, S. Peters, L. R. Teras, E. R. Watkins, J. L. Burgess, A. A. Stec, P. A. White, N. L. DeBono, L. Benbrahim-Tallaa, A. D. Conti, F. E. Ghissassi, Y. Grosse, L. T. Stayner, E. Suonio, S. Viegas, R. Wedekind, P. Boucheron, B. Hosseini, J. Kim, H. Zahed, H. Mattock, F. Madia and M. Schubauer-Berigan, *Lancet Oncol.*, 2022, **23**, 985–986.
3. J. M. Goodrich, M. M. Calkins, A. Caban-Martinez, T. Stueckle, C. Grant, A. M. Calafat, A. Nematollahi, A. M. Jung, J. M. Graber, T. Jenkins, A. L. Slitt, A. Dewald, J. C. Botelho, S. Beitel, S. Littau, J. Gulotta, D. Wallentine, J. Hughes, C. Popp and J. L. Burgess, *Epigenomics*, 2021, **13**, 1619–1636.
4. Y. Su, J. Li and G. Song, *Int. J. Occup. Saf. Ergon.*, 2018, **24**, 190–199.
5. Why Firefighters Get Steam Burns: Exploratory Study Underway, https://www.nist.gov/news-events/news/2015/04/why-firefighters-get-steam-burns-exploratory-study-underway#:~:text=In%20the%20potential%20steam%20burn,humidity%20in%20this%20air%20layer.
6. D. J. Muensterman, I. A. Titaley, G. F. Peaslee, L. D. Minc, L. Cahuas, A. E. Rodowa, Y. Horiuchi, S. Yamane, T. N. J. Fouquet, J. C. Kissel, C. C. Carignan and J. A. Field, *Environ. Sci. Technol.*, 2022, **56**, 974–983.
7. J. A. Laitinen, J. Koponen, J. Koikkalainen and H. Kiviranta, *Toxicol. Lett.*, 2014, **231**, 227–232.
8. J. Trowbridge, R. R. Gerona, T. Lin, R. A. Rudel, V. Bessonneau, H. Buren and R. Morello-Frosch, *Environ. Sci. Technol.*, 2020, **54**, 3363–3374.
9. R. D. Daniels, T. L. Kubale, J. H. Yiin, M. M. Dahm, T. R. Hales, D. Baris, S. H. Zahm, J. J. Beaumont, K. M. Waters and L. E. Pinkerton, *Occup. Environ. Med.*, 2014, **71**, 388.
10. N. L. Wagner, J. Berger, D. Flesch-Janys, P. Koch, A. Köchel, M. Peschke and T. Ossenbach, *Environ. Health*, 2006, **5**, 27.
11. N. F. P. Association, *NFPA 1971: Standard on Protective Ensembles for Structural Firefighting and Proximity Firefighting*, 2018.
12. About NFPA, https://www.nfpa.org/About-NFPA.

13. Latest action on firefighting PPE standard underscores the need to better understand and to participate in the standards process, https://www.nfpa.org/News-and-Research/Publications-and-media/Blogs-Landing-Page/NFPA-Today/Blog-Posts/2021/11/05/Latest-action-on-firefighting-PPE-standard-underscores-the-need-to-better-understand.
14. First draft of NFPA 1970 proposes changes to firefighter PPE standard and will be open for review and Public Comment through January 4, 2023, https://www.nfpa.org/News-and-Research/Publications-and-media/Blogs-Landing-Page/NFPA-Today/Blog-Posts/2023/04/04/NFPA-releases-Standards-Council-decision-on-NFPA-1582-addressing-possible-legal.
15. Greener Solutions 2022: PFAS-free Fire Fighting Turnout Gear, https://bcgc.berkeley.edu/greener-solutions-2022-pfas-free-fire-fighting-turnout-gear.
16. GORE® PARALLON™ Liner System, https://us.msasafety.com/products/firefighter-safety/materials/moisture-barrier/gore-parallon.
17. V. Mokoana, J. Asante and J. Okonkwo, *J. Fire Sci.*, 2021, **39**, 207–223.
18. K.-Y. Law, *J. Phys. Chem. Lett.*, 2014, **5**, 686–688.
19. Insulation Basics, https://local.armacell.com/fileadmin/cms/singapore/Campaign/Armacell_KnowHow-1-Important_Insulation_Terms.pdf.
20. D. Lin, X. Zeng, H. Li, X. Lai and T. Wu, *J. Colloid Interface Sci.*, 2019, **533**, 198–206.
21. GreenScreen for safer chemicals: Hazard assessment guidance, https://www.greenscreenchemicals.org/images/ee_images/uploads/resources/GreenScreen_Guidance_v1_4_2018_01_Final.pdf
22. REACH – registration, evaluation, authorisation and restriction of chemicals regulation. European Chemical Agency, https://echa.europa.eu/information-on-chemicals/registered-substances.
23. PFAS and Fire Fighter Turnout Gear, https://www.iaff.org/pfas/
24. Y. N. Kim, Y.-M. Ha, J. E. Park, Y.-O. Kim, J. Y. Jo, H. Han, D. C. Lee, J. Kim and Y. C. Jung, *Polym. Test.*, 2021, **93**, 107006.
25. C. Felgueiras, N. G. Azoia, C. Gonçalves, M. Gama and F. Dourado, *Front. Bioeng. Biotechnol.*, 2021, **9**, 608826.
26. Cellulose Nanofiber Manufacturing Technology and Application Development, https://www.nipponpapergroup.com/english/research/organize/cnf.html.
27. Lyocell Vs. Cotton Sheets: Which is Best?, https://www.nipponpapergroup.com/english/research/organize/cnf.html.
28. K. Moriam, M. Rissanen, D. Sawada, M. Altgen, L.-S. Johansson, D. V. Evtyugin, C. Guizani, M. Hummel and H. Sixta, *ACS Sustainable Chem. Eng.*, 2021, **9**, 4915–4925.
29. P. Manik, S. Majumder and R. K. Hossain, *Trends Text. Eng. Fashion Technol.*, 2019, **5**, 583.
30. Z. Li, J. Yang and X. J. Loh, *NPG Asia Mater.*, 2016, **8**, e265.
31. S. Kopf, D. Åkesson and M. Skrifvars, *Polym. Rev.*, 2023, **63**, 200–245.

32. E. E. Tănase, M. E. Popa, M. Râpă and O. Popa, *Agric. Agric. Sci. Procedia*, 2015, **6**, 608–615.
33. H. Abdali and A. Ajji, in *Carbon Nanostructures*, ed. A. Khan, M. Jawaid, B. Neppolian and A. M. Asiri, Springer Singapore, Singapore, 2019, pp. 127–156.
34. A. Al-Dhahebi, S. C. B. Gopinath and M. S. M. Saheed, *Nano Convergence*, 2020, **7**, 27.
35. What Factors Impact Graphene Cost, https://investingnews.com/daily/tech-investing/nanoscience-investing/graphene-investing/graphene-cost/
36. Graphene typically costs $200 000 per ton. Now, scientists can make it from trash, https://bigthink.com/the-present/flash-graphene/#:~:text=%24200%2C000%20per%20ton.-,Now%2C%20scientists%20can%20make%20it%20from%20trash.,difficult%20to%20produce%20%E2%80%94%20until%20now.&text=The%20new%20technique%2C%20developed%20at,material%20in%20just%2010%20milliseconds.
37. Association of Environmental and Occupational Clinics (AEOC), Assessment search: Cellulose.
38. S. Wang, W. Chen, H. Xiang, J. Yang, Z. Zhou and M. Zhu, *Polymers*, 2016, **8**, 273.
39. 4-Oxo-4-[(4-oxobutan-2-yl)oxy]butan-2-yl 3-hydroxybutanoate, CompTox Chemicals Dashboard v2.2, https://comptox.epa.gov/dashboard/chemical/executive-summary/DTXSID30951955.
40. European Chemicals Agency (ECHA), Assessment Search: Graphene, https://echa.europa.eu/substance-information/-/substanceinfo/100.227.924
41. S. Hamdani, C. Longuet, D. Perrin, J. Lopez-cuesta and F. Ganachaud, *Polym. Degrad. Stab.*, 2009, **94**, 465–495.
42. A. D. Naik, G. Fontaine, S. Bellayer and S. Bourbigot, *RSC Adv.*, 2015, **5**, 48224–48235.
43. N. Li, J. Ming, R. Yuan, S. Fan, L. Liu, F. Li, X. Wang, J. Yu and D. Wu, *ACS Sustainable Chem. Eng.*, 2020, **8**, 290–301.
44. S. Hörold, *Polymer Green Flame Retardants*, ed. C. D. Papaspyrides and P. Kiliaris, Elsevier, Amsterdam, 2014, pp. 221–254.
45. I. V. D. Veen and J. D. Boer, *Chemosphere*, 2012, **88**, 1119–1153.
46. Z. Zhang, C. Dong, J. Liu, D. Kong, L. Sun and Z. Lu, *Cellulose*, 2020, **27**, 1799–1815.
47. Exolit® 5060 PK, https://www.clariant.com/en/Solutions/Products/2014/03/18/16/31/Exolit-5060-PK.
48. European Chemicals Agency (ECHA), Assessment Search: 9,10-dihydro-9-oxa-10-phosphaphenanthrene-10-oxide, https://echa.europa.eu/substance-information/-/substanceinfo/100.047.998.
49. Ministry of Environment Denmark, Environmental Protection Agency (Danish EPA), 9,10-dihydro-9-oxa-10-phosphaphenanthrene 10-Oxide, https://eng.mst.dk/chemicals/chemicals-in-products/assessment-of-chemicals/the-advisory-list-for-self-classification-of-hazardous-substances/.

50. Ministry of Environment Denmark, Environmental Protection Agency (Danish EPA), 1,3-Bis(3-aminopropyl)-1,1,3,3-tetramethyldisiloxane, https://eng.mst.dk/chemicals/chemicals-in-products/assessment-of-chemicals/the-advisory-list-for-self-classification-of-hazardous-substances/.
51. Japan GHS Regulation, Assessment Search: Tetraethyl silicate, https://www.nite.go.jp/chem/english/ghs/ghs_index.html.
52. New Zealand GHS Regulation, Assessment Search: Polydimethylsiloxane, https://www.epa.govt.nz/database-search/chemical-classification-and-information-database-ccid/.
53. 1,3,2-Dioxaphosphorinane, 2,2′-oxybis[5,5-dimethyl-, 2,2′-disulfide. CompTox Chemicals Dashboard v2.2, https://comptox.epa.gov/dashboard/chemical/details/DTXSID4063290.
54. European Chemicals Agency (ECHA), Assessment Search: 1,3,2-Dioxaphosphorinane, 2,2′-oxybis[5,5-dimethyl-, 2,2′-disulfide, https://echa.europa.eu/substance-information/-/substanceinfo/100.021.663.
55. IARC Monographs on the evaluation of Carcinogenic Risks to Humans, Assessment Search: Diazinon, https://www.ncbi.nlm.nih.gov/books/NBK419324/#_NBK419324_pubdet_.
56. Japan GHS Regulation, Assessment Search: Diazinon, https://www.nite.go.jp/chem/english/ghs/ghs_index.html.
57. New Zealand GHS Regulation, Assessment Search: Diazinon, https://www.epa.govt.nz/database-search/chemical-classification-and-information-database-ccid/.
58. Malaysia GHS Regulation, Assessment Search: Diazinon, https://www.dosh.gov.my/index.php/legislation/codes-of-practice/chemical-management.
59. Collaborative on Health and the Environment (CHE), Diazinon, https://www.healthandenvironment.org/our-work/toxicant-and-disease-database/?showcategory=&showdisease=&showcontaminant=&showcas=333-41-5&showkeyword=.
60. European Union (EU) Priority Endocrine Disruptors, Assessment search: Diazinon, https://commission.europa.eu/strategy-and-policy/policies/endocrine-disruptors_en#priority_list.
61. Oregon Department of Environmental Quality (Oregon DEQ), Assessment search: Diazinon, https://web.archive.org/web/20150912220325/http://www.deq.state.or.us/wq/SB737/docs/LegRpAtt20100601.pdf.
62. European Union (EU) GHS Regulation (H-Statements) Annex 6 Table 3-1, Assessment Search: Diazinon, https://echa.europa.eu/web/guest/information-on-chemicals/cl-inventory-database.
63. P. Grandjean and P. J. Landrigan, *Lancet Neurol.*, 2014, **13**, 648–649.
64. California Office of Environmental Health Hazard Assessment (OEHHA), The Proposition 65 List, Chlorpyrifos, https://oehha.ca.gov/proposition-65/chemicals/chlorpyrifos.
65. Japan GHS Regulation, Assessment Search: Chlorpyrifos, https://www.nite.go.jp/chem/english/ghs/ghs_index.html.

66. Oregon Department of Environmental Quality (Oregon DEQ), Assessment search: Chlorpyrifos, https://web.archive.org/web/20150912220325/http://www.deq.state.or.us/wq/SB737/docs/LegRpAtt20100601.pdf.
67. S. Ebnesajjad, in *Expanded PTFE Applications Handbook*, ed. S. Ebnesajjad, William Andrew Publishing, Oxford, 2017, pp. 163–170.
68. M. Davoudi, E. J. Moore and G. G. Chase, *J. Taiwan Inst. Chem. Eng.*, 2019, **94**, 24–30.
69. Introducing the Expanded Polyethylene (EPE) Membrane, https://www.gore-tex.com/en_uk/blog/introducing-the-expanded-polyethylene-epe-membrane-0.
70. C. Vasile and M. Pascu, *Practical guide to polyethylene*, iSmithers Rapra Publishing, 2005.
71. C. Akduman and E. P. A. Kumbasar, *Aspects of Polyurethanes*, InTech, 2017.
72. T.-T. Li, P. Liu, H. Wang, W. Dai, J. Wang, Z. Wang, B.-C. Shiu, C.-W. Lou and J.-H. Lin, *J. Mater. Res. Technol.*, 2021, **12**, 1288–1302.
73. Association of Environmental and Occupational Clinics (AEOC), Assessment search: Polyethylene, http://www.aoecdata.org/ExpCodeLookup.aspx.
74. Organisation for Economic Co-operation and Development (OECD), Assessment search: Polyethylene, https://canadachemicals.oecd.org/ChemicalDetails.aspx?Key=b89f4f0e-39c2-4caf-9287-75b4464c7be5&Idx=0.
75. Permanent Senate Commission for the Investigation of Health Hazards of Chemical Compounds in the Work Area (MAK-Commission of Germany), Assessment search: Ethylene, https://onlinelibrary.wiley.com/doi/book/10.1002/3527600418.
76. European Chemicals Agency (ECHA), Ethylene, https://echa.europa.eu/substance-information/-/substanceinfo/100.000.742.
77. European Union (EU) GHS Regulation (H-Statements) Annex 6 Table 3-1, Assessment Search: Ethylene, https://echa.europa.eu/web/guest/information-on-chemicals/cl-inventory-database.
78. Polyurethane foam. CompTox Chemicals Dashboard v2.2, https://comptox.epa.gov/dashboard/chemical/hazard/DTXSID201031480.
79. European Chemicals Agency (ECHA), Glycerol, https://echa.europa.eu/substance-information/-/substanceinfo/100.000.263.
80. European Chemicals Agency (ECHA), 4,4′-Methylenediphenyl diisocyanate, https://echa.europa.eu/substance-information/-/substanceinfo/100.002.697.
81. Peeling back the fire suit "onion", layer #2: The moisture barrier, https://eu.tencatefabrics.com/blog/fire-suit-layer-moisture-barrier.
82. E. Standards, *Protective Clothing for Firefighters – Performance requirements for protective clothing for firefighter activities. BSI Standards Limited – EN 469:2020.*
83. S. A. Limited, *Protective clothing for firefighters — Requirements and test methods for protective clothing used for structural firefighting*, 2019.

84. Canadian Environmental Protection Act (CEPA), Categorization Results from the Canadian Domestic Substance List, Assessment Search: Polytetrafluoroethylene, https://canadachemicals.oecd.org/Search.asp.
85. Québec Workplace Health and Safety Commission (Commission de la santé et de la securite du travail (CSST)), Assessment Search: Tetrafluoroethylene, https://reptox.cnesst.gouv.qc.ca/Pages/repertoire-toxicologique.aspx.
86. Permanent Senate Commission for the Investigation of Health Hazards of Chemical Compounds in the Work Area (MAK-Commission of Germany), Assessment search: Tetraethyl silicate, https://onlinelibrary.wiley.com/doi/book/10.1002/3527600418.
87. California Office of Environmental Health Hazard Assessment (OEHHA), The Proposition 65 List, Tetrafluoroethylene, https://oehha.ca.gov/proposition-65/chemicals/tetrafluoroethylene.
88. Japan GHS Regulation, Assessment Search: Tetrafluoroethylene, https://www.nite.go.jp/chem/english/ghs/ghs_index.html.
89. European Union (EU) GHS Regulation (H-Statements) Annex 6 Table 3-1, Assessment Search: Perfluoroctonanoic acid, https://echa.europa.eu/web/guest/information-on-chemicals/cl-inventory-database.
90. Japan GHS Regulation, Assessment Search: Perfluoroctonanoic acid, https://www.nite.go.jp/chem/english/ghs/ghs_index.html.
91. Canadian Environmental Protection Act (CEPA), Categorization Results from the Canadian Domestic Substance List, Assessment Search: Perfluoroctonanoic acid, https://canadachemicals.oecd.org/Search.asp.
92. California Office of Environmental Health Hazard Assessment (OEHHA), The Proposition 65 List, Perfluoroctonanoic acid, https://oehha.ca.gov/proposition-65/chemicals/perfluorooctanoic-acid-pfoa-and-its-salts.
93. TEDX, TEDX (The Endocrine Disruption eXchange) List of Potential Endocrine Disruptors, https://endocrinedisruption.org/interactive-tools/tedx-list-of-potential-endocrine-disruptors/search-the-tedx-list.
94. ChemSec, The International Chemical Secretariat, Substitute It Now (SIN) List, List Search: Perfluorooctanesulfonic acid, http://sinlist.chemsec.org/.
95. Oregon Department of Environmental Quality (Oregon DEQ), Assessment search: Perfluorooctanesulfonic acid, https://web.archive.org/web/20150912220325/http://www.deq.state.or.us/wq/SB737/docs/LegRpAtt20100601.pdf.
96. European Union (EU) GHS Regulation (H-Statements) Annex 6 Table 3-1, Assessment Search: Perfluorooctanesulfonic acid, https://echa.europa.eu/web/guest/information-on-chemicals/cl-inventory-database.
97. California Office of Environmental Health Hazard Assessment (OEHHA), The Proposition 65 List, Perfluorooctane sulfonate, https://oehha.ca.gov/proposition-65/chemicals/perfluorooctane-sulfonate-pfos

98. Japan GHS Regulation, Assessment Search: Perfluorooctanesulfonic acid, https://www.nite.go.jp/chem/english/ghs/ghs_index.html.
99. 2,3,3,3-Tetrafluoro-2-(heptafluoropropoxy)propionic acid. CompTox Chemicals Dashboard v2.2, https://comptox.epa.gov/dashboard/chemical/details/DTXSID70880215.
100. European Chemicals Agency (ECHA), 2,3,3,3-Tetrafluoro-2-(heptafluoropropoxy)propionic acid, https://echa.europa.eu/substance-information/-/substanceinfo/100.032.928.
101. European Union (EU) GHS Regulation (H-Statements) Annex 6 Table 3-1, Assessment Search: perfluorodecanoic acid, https://echa.europa.eu/web/guest/information-on-chemicals/cl-inventory-database.
102. ChemSec, The International Chemical Secretariat, Substitute It Now (SIN) List, List Search: Perfluorodecanoic acid, http://sinlist.chemsec.org/.
103. Perfluorodecanoic acid, CompTox Chemicals Dashboard v2.2, https://comptox.epa.gov/dashboard/chemical/details/DTXSID3031860.
104. ChemSec, The International Chemical Secretariat, Substitute It Now (SIN) List, List Search: Tetrabromobisphenol A, http://sinlist.chemsec.org/.
105. California Office of Environmental Health Hazard Assessment (OEHHA), The Proposition 65 List, Tetrabromobisphenol A, https://oehha.ca.gov/proposition-65/chemicals/tetrabromobisphenol.
106. Canadian Environmental Protection Act (CEPA), Categorization Results from the Canadian Domestic Substance List, Assessment Search: Tetrabromobisphenol A, https://canadachemicals.oecd.org/Search.asp.
107. Japan GHS Regulation, Assessment Search: Tetrabromobisphenol A, https://www.nite.go.jp/chem/english/ghs/ghs_index.html.
108. B. J. Carroll, *J. Colloid Interface Sci.*, 1976, **57**, 488–495.
109. R. F. Mahbub, L. Wang, L. Arnold, S. Kaneslingam and R. Padhye, *Text. Res. J.*, 2014, **84**, 2094–2102.
110. All About Kevlar, https://www.fiberbrokers.com/body-armor-disposal/all-about-kevlar/
111. S. K. Li, R. P. Smith and A. W. Neumann, *J. Adhes.*, 1984, **17**, 105–122.
112. M. Yang, Z. Zhang, J. Yuan, L. Wu, X. Zhao, F. Guo, X. Men and W. Liu, *Friction*, 2020, **8**, 335–342.
113. DuPont™ Nomex® 410 Paper, 20 mil Nominal Thickness.
114. F. Taherian, V. Marcon, N. F. A. van der Vegt and F. Leroy, *Langmuir*, 2013, **29**, 1457–1465.
115. L. Li, B. Zhou, G. Han, Y. Feng, C. He, F. Su, J. Ma and C. Liu, *Compos. Sci. Technol.*, 2020, **197**, 108229.
116. N. Jamaluddin, T. Kanno, T.-A. Asoh and H. Uyama, *Mater. Today Commun.*, 2019, **21**, 100587.
117. C. D. Midhun Dominic, R. Joseph, P. M. S. Begum, M. Joseph, D. Padmanabhan, L. A. Morris, A. S. Kumar and K. Formela, *Polymers*, 2020, **12**, 814.

118. C. Ahn, H. J. Yoo, Y.-S. Oh, S.-S. Han, H. J. Lee, J. H. Kim, K.-H. Song and J. S. Rhie, *Text. Res. J.*, 2005, **75**, 139–143.
119. A. Basit, W. Latif, M. Ashraf, A. Rehman, K. Iqbal, S. M. Hafiz, A. Jabbar and A. B. Sajjad, *Autex Res. J.*, 2019, **19**, 80–85.
120. A. Gericke and J. V. D. Pol, *J. Fam. Ecol. Consum. Sci.*, 2010, **38**, 63.
121. Y. L. Hsieh, *Text. Res. J.*, 1994, **64**, 552–553.
122. O. Nechyporchuk, R. Bordes and T. Köhnke, *ACS Appl. Mater. Interfaces*, 2017, **9**, 39069–39077.
123. S. Hosseini, P. Azari, E. Farahmand, S. N. Gan, H. A. Rothan, R. Yusof, L. H. Koole, I. Djordjevic and F. Ibrahim, *Biosens. Bioelectron.*, 2015, **69**, 257–264.
124. S. G. Lee, S.-S. Choi, W. H. Park and D. Cho, *Macromol. Symp.*, 2003, **197**, 89.
125. Z. Liu, J. Lyu, D. Fang and X. Zhang, *ACS Nano*, 2019, **13**, 5703–5711.
126. D. Serbezeanu, A. M. Popa, T. Stelzig, I. Sava, R. M. Rossi and G. Fortunato, *Text. Res. J.*, 2015, **85**, 1763–1775.
127. Y. Sun, W. Chen and X. Zhou, *RSC Adv.*, 2021, **11**, 34828–34835.
128. P. Kofler, A. Herten, D. Heinrich, G. Bottoni, M. Hasler, M. Faulhaber, T. Bechtold, W. Nachbauer and M. Burtscher, *Text. Res. J.*, 2013, **83**, 2085–2092.
129. L. Yang, J. Z. Yang and L. Li, *Adv. Mater. Res.*, 2014, **1048**, 41–45.
130. C. Cui, W. Qian, Y. Yu, C. Kong, B. Yu, L. Xiang and F. Wei, *J. Am. Chem. Soc.*, 2014, **136**, 2256–2259.
131. C. Gong, Y. He, J. Zhou, W. Chen, W. Han, Z. Zhang, P. Zhang, X. Pan, Z. Wang and E. Xie, *ACS Appl. Mater. Interfaces*, 2014, **6**, 14844–14850.
132. K. Shanmugam, N. Chandrasekar and R. Balaji, *Micro*, 2023, **3**, 207.
133. H. Ma, C. Burger, B. S. Hsiao and B. Chu, *J. Mater. Chem.*, 2011, **21**, 7507–7510.
134. H. Ozdemir, *J. Eng. Fibers Fabr.*, 2017, **12**, 155892501701200102.
135. I. S. Cubric, Z. Skenderi and G. Havenith, *Text. Res. J.*, 2013, **83**, 1215–1228.
136. The Functional Properties of "Tencel", https://www.proquest.com/newspapers/functional-properties-tencel/docview/233350639/se-2?accountid=14496 https://search.library.berkeley.edu/openurl/01UCS_BER/01UCS_BER:UCB?url_ver=Z39.88-2004&rft_val_fmt=info:ofi/fmt:kev:mtx:journal&genre=article&sid=ProQ:ProQ%253Amaterialsscijournals&atitle=The+Functional+Properties+of+%2526quot%253BTencel%2526quot%253B&title=Asian+Textile+Business&issn=&date=2005-07-01&volume=&issue=608&spage=47&au=Firgo%252C+Heinrich&isbn=&jtitle=Asian+Textile+Business&btitle=&rft_id=info:eric/&rft_id=info:doi/.
137. S. S. Ramkumar, A. Purushothaman, K. D. Hake and D. D. McAlister, *J. Eng. Fibers Fabr.*, 2007, **2**, 155892500700200.
138. A. Ivanovska, M. Reljic, M. Kostic, K. Asanovic and B. Mangovska, *J. Nat. Fibers*, 2022, **19**, 5465–5477.
139. S.-W. Peng, X.-Y. Guo, G.-G. Shang, J. Li, X.-Y. Xu, M.-L. You, P. Li and G.-Q. Chen, *Biomaterials*, 2011, **32**, 2546–2555.

140. New Zealand GHS Regulation, Assessment Search: 3,4-Dihydroxybenzaldehyde, https://www.epa.govt.nz/database-search/chemical-classification-and-information-database-ccid/.
141. European Chemicals Agency (ECHA), 1,3-Bis(3-aminopropyl)-1,1,3,3-tetramethyldisiloxane, https://echa.europa.eu/substance-information/-/substanceinfo/100.017.808.
142. German Federal Environment Agency (FEA), Substances Hazardous to Waters, Assessment Search: Tetraethyl silicate, http://webrigoletto.uba.de/Rigoletto/.
143. European Union (EU) GHS Regulation (H-Statements) Annex 6 Table 3-1, Assessment Search: Tetraethyl silicate, https://echa.europa.eu/web/guest/information-on-chemicals/cl-inventory-database.
144. German Federal Environment Agency (FEA), Substances Hazardous to Waters, Assessment Search: Polydimethylsiloxane, http://webrigoletto.uba.de/Rigoletto/.
145. ChemForward, Assessment Search: 1,3,2-Dioxaphosphorinane, 2,2′-oxybis[5,5-dimethyl-, 2,2′-disulfide, https://alternatives.chemforward.org/preview/a6nbmi363e6a/portfolios/17.
146. New Zealand GHS Regulation, Assessment Search: Chlorpyrifos, https://www.epa.govt.nz/database-search/chemical-classification-and-information-database-ccid/.
147. European Chemicals Agency (ECHA), Chlorpyrifos, https://echa.europa.eu/substance-information/-/substanceinfo/100.018.969.
148. K. S. Kim and C. H. Park, *Text. Res. J.*, 2013, **83**, 1808–1820.
149. L. Chen, H. He, M. Wang, X. Li and H. Yin, *Tissue Eng. Regener. Med.*, 2017, **14**, 359–370.
150. P. Kuang and K. Constant, Increased Wettability and Surface Free Energy of Polyurethane by Ultraviolet Ozone Treatment, in *Wetting and Wettability*, ed. M. Aliofkhazraei, IntechOpen, Rijeka, 2015, ch. 4.
151. M. A. Jahid and J. Hu, *Mater. Today: Proc.*, 2019, **16**, 1485–1490.
152. Y. K. Kang, C. H. Park, J. Kim and T. J. Kang, *Fibers Polym.*, 2007, **8**, 564–570.
153. M. A. Aboamer, A. A. Aboamer, D. H. Elgohary, T. M. Alqahtani, A. Abdel-Hadi, S. M. Al-Mutairi, T. M. El-Bagory, K. M. Alshareef and N. A. R. Mohamed, *J. Mech. Sci. Technol.*, 2022, **36**, 5977–5993.
154. E. Lock, S. Walton and R. Fernsler, *Preparation of Ultra Thin Polystyrene, Polypropylene and Polyethylene Films on Si Substrate Using Spin Coating Technology*, Naval Research Lab Washington Dc Plasma Physics Div, 2008.
155. Hydrophobic ePTFE Membrane, https://www.cobetterfiltration.com/Product/Hydrophobic_Oleophobic_PTFE_Membrane.html#:~:text=The%20hydrophobic%20nature%20of%20PTFE,always%20effective%20while%20maintaining%20the.
156. ePTFE Waterproof and moisture permeable composite fabric wholesale, https://www.eptfemembrane.net/product/eptfe-waterproof-and-moisture-permeable-composite-fabric.html.
157. X. Liu, W. Xu, W. Li, Y. Chen and J. Rao, *Polym. Eng. Sci.*, 2010, **50**, 2400–2407.

158. S.-i. Aiba, N. Minoura, Y. Fujiwara, S. Yamada and T. Nakagawa, *Biomaterials*, 1985, **6**, 290–296.
159. H. Haugen, V. Ried, M. Brunner, J. Will and E. Wintermantel, *J. Mater. Sci.: Mater. Med.*, 2004, **15**, 343–346.
160. Sintered Polyethylene, https://www.porex.com/porous-polymers-technology/sintered-porous-plastic/sintered-polyethylene-2/.
161. E. Hatfield, R. Tate, K. Williams and W. Todd, *J. Plast. Film Sheeting*, 2002, **18**, 117–127.
162. D. M. Seong, H. Lee, J. Kim and J. H. Chang, *Materials*, 2020, **13**, 2382.
163. IARC Monographs on the evaluation of Carcinogenic Risks to Humans, Assessment Search: Polyethylene, https://www.ncbi.nlm.nih.gov/books/NBK419324/#_NBK419324_pubdet_.
164. Dow Product Information, VORANOLTM 3010 Polyether Polyol, https://www.dow.com/content/dam/dcc/documents/en-us/productdatasheet/109/109-01450-01-voranol-3010-polyether-polyol-tds.pdf.
165. Polyurethane, CompTox Chemicals Dashboard v2.2, https://comptox.epa.gov/dashboard/chemical/details/DTXSID5031929.
166. Japan GHS Regulation, Assessment Search: Ethylene, https://www.nite.go.jp/chem/english/ghs/ghs_index.html.

CHAPTER 10

Is a PFAS-free Future Possible?

S. A. BĂLAN,*[a,b] T. A. BRUTON,[a] C. LEONETTI[a] AND K. G. HAZARD[b]

[a] California Department of Toxic Substances Control, Sacramento, CA, 95812, USA; [b] University of California, Berkeley, School of Public Health, Berkeley, CA, 94704, USA
*Email: simona.balan@dtsc.ca.gov

10.1 Asking the Big Question

Perfluoroalkyl and polyfluoroalkyl substances (PFASs) have been used in a wide range of consumer products and industrial applications for decades, enabling unique functionalities, elevating societal standards of living, and generating profits. So perhaps one might wonder: Is a PFAS-free future possible? On February 7, 2023, while wrapping up this book, we asked this question to ChatGPT, an artificial intelligence language model.[†] This was its answer (shortened for relevance):

> "*A PFAS-free future may not be fully achievable, as PFASs have been widely used in industrial and consumer products for several decades (…). However, a significant reduction in the use and exposure to PFASs is possible through a combination of regulatory action and technological innovation. (…) It is likely that a phased approach, with a focus on reducing exposure to the most harmful PFASs and promoting the use of safer alternatives, will be necessary in order to minimize the impact of PFASs on public health and the environment.*"

[†]ChatGPT is a product of OpenAI (https://chat.openai.com/).

Green Chemistry Series No. 81
Toward a PFAS-free Future: Safer Alternatives to Forever Chemicals
Edited by Simona A. Bălan, Thomas A. Bruton and Kimberly G. Hazard

Published by the Royal Society of Chemistry, www.rsc.org

Coincidentally, that same day, something big happened: the European Chemicals Agency (ECHA) published its nearly 2000-page long PFAS restriction proposal, prepared by the authorities in Denmark, Germany, the Netherlands, Norway, and Sweden.[1] If approved, this regulation would ban virtually all uses of PFASs in the European Union, with some time-bound exemptions.

When asked that same question again three days later, ChatGPT responded:

> *"A PFAS-free future is challenging, but possible with significant effort and dedication. (...) To achieve a PFAS-free future, several steps must be taken, including research to identify alternative, safer chemicals and materials, regulations to limit the use of PFAS, and increased public awareness and demand for PFAS-free products. It is important to note that a transition to a PFAS-free future will not happen overnight and will require a collaborative effort from various sectors, including government, industry, and the public. But with determination and a focus on creating a healthier, safer, and more sustainable future, it is possible to reduce and ultimately eliminate the use of PFAS."*

The change in ChatGPT's answer is unrelated to ECHA's PFAS ban proposal (its training data only goes up to 2021). In fact, after another three days (February 13, 2023), ChatGPT changed its answer yet again:

> *"A completely PFAS-free future may be difficult to achieve, as PFASs have been used in a wide range of products for many years and are highly persistent in the environment. However, it is possible to reduce the use and release of PFASs and to transition towards more sustainable alternatives. (...) While a completely PFAS-free future may not be immediately achievable, it is important to continue to work towards reducing the use and release of these substances in order to protect the environment and human health."*

Moral of the story? It is complicated.

And yet, it is also fairly straightforward. As demonstrated in previous chapters, small teams of university students can, in the span of one semester, use the principles of green chemistry coupled with nature's infinite source of inspiration to identify safer bio-based and biodegradable alternatives to PFASs in consumer products. Whether they were tasked with identifying safer and functionally acceptable PFAS alternatives for food packaging, household goods packaging, textiles, or floor polish, or strategies for removing PFASs from recycled carpet fiber, the student teams successfully found candidates that met their performance criteria and had improved safety profiles. The complexity arises in translating such ideas into practice in the real world. With many societal changes, the main obstacle in advancing toward a PFAS-free future is the strong inertia that pulls us in the opposite direction.

The case study from the 2022 cohort of the Greener Solutions course at the University of California, Berkeley, featured in Chapter 9 illustrates the complexity of overcoming this inertia. Partnering with the International Association of Firefighters (IAFF), the team sought safer alternatives for polytetrafluoroethylene (PTFE) moisture barriers in firefighter turnout gear. They found that prescriptive standards, such as light degradation and viral resistance, are unnecessary and hinder the adoption of PFAS-free options. As this student team found first-hand, the path toward a PFAS-free future may be hindered by both policy and technical barriers.

The journey toward a PFAS-free future will require not only bright ideas and out-of-the-box thinking, but also a paradigm shift in chemical manufacturing and use, as well as large scale collaborations among industry, academia, government, and public interest groups. Gathering enough momentum toward safer PFAS alternatives on a societal scale requires a right combination of factors. This chapter surveys three interrelated types of factors that have been pushing the market toward a PFAS-free future: public and political pressure, legislative and regulatory pressure, and supply chain pressure. We discuss their impacts to date and conclude with some recommendations for green chemists and industrial chemists seeking to identify safer alternatives to PFASs.

10.2 Public and Political Pressure

Contamination of drinking water supplies has, in major part, been responsible for the high level of public concern over PFASs. As mentioned in Chapter 2, the discovery of drinking water contamination in the Ohio River Valley stemming from DuPont's use of perfluorooctanoic acid (PFOA)[2] was a turning point. Another important milestone was the first identification, around the year 2000, of PFAS groundwater contamination associated with the use of aqueous film-forming foams (AFFFs).[3] Together, these two events accelerated the scientific research into the environmental fate and transport, human exposure routes, toxicokinetics, and toxicity of PFASs.

As the scientific community's understanding of PFASs grew, so did its concerns. Publication of the Helsingor[4] and Madrid[5] Statements in 2014 and 2015, respectively, reflected the emerging consensus among environmental scientists that the unique challenges posed by this class of chemicals merited swift action to mitigate harm. These consensus statements, which called for limiting the use of PFASs in favor of non-persistent alternatives, resulted in significant media coverage and increased awareness of PFASs among policy makers, industry, and the public.

A 2016 analysis[6] of EPA data showing that drinking water supplies for six million U.S. residents contained PFOA and perfluorooctane sulfonate (PFOS) in excess of EPA's Lifetime Health Advisory also garnered significant media attention. Around the same time, a wave of discoveries of PFAS-contaminated water supplies in New Hampshire, New York, Vermont, Michigan, North Carolina, Maine, and other states gave rise to grassroots groups that advocated for drinking water treatment, biomonitoring, and polluter accountability. These community-based groups were effective in garnering local and national media coverage and succeeded in bringing the issue of PFAS-contaminated drinking water to the attention of state and federal legislators and environmental regulators.

By 2019, there was even an official PFAS caucus in the U.S. House of Representatives, consisting mostly of representatives whose districts include communities with PFAS-contaminated water. Legislators in the U.S. House of Representatives and Senate have proposed numerous pieces of PFAS-related legislation in the past five years. Many of these bills have focused on addressing drinking water contamination and limiting the use of PFAS-containing firefighting foam, but over time the focus has widened to include proposals for reducing PFAS use in other types of products.[7–9]

In 2020, the Hollywood film *Dark Waters* brought the story of PFOA contamination in the Ohio River valley to millions of viewers.[10] The same year, proposals for addressing PFAS contamination were featured in campaign materials for the Democratic party's candidate in the U.S. presidential election.[11] The public's and policymakers' increasing concern over PFASs has been widely reported in the business press, strongly signalling to manufacturers that there is a need to identify safer alternative chemicals.

10.3 Legislative and Regulatory Pressure

As of this writing, eleven U.S. states have enacted legislation that bans the use of PFASs in certain consumer products, including food packaging, carpets and other home textiles, aftermarket textile treatments, children's products, cosmetics, and ski wax (refer to Chapter 2 for details). U.S. states including California and Washington have passed new regulations restricting some consumer product uses of PFASs or requiring manufacturers of those products to search for safer alternatives. U.S. EPA has proposed a regulation requiring manufacturers and importers of PFAS-containing articles to report on the identities and amounts of PFAS used.

These legislative bans and regulatory restrictions will likely lead to the replacement of PFASs will safer alternatives in these product categories. At the same time, the number of product categories subject to these bans and restrictions is still relatively small, and many industries continue the unregulated use of PFAS use (*e.g.*, semiconductors, electronics, renewable energy components, and building products).

The lack of a widely accepted PFAS definition has resulted in inconsistent language in various laws and regulations, which makes it difficult for businesses to comply. Regulators also struggle with enforcing PFAS restrictions in products due to a lack of analytical methods for detecting most members of the class.

Nevertheless, this increasing regulatory pressure is one of the main factors moving certain consumer product segments away from PFASs. Maine's ban on all non-essential uses of PFASs and EU's proposal to restrict virtually all uses of PFASs under REACH, with some time-bound exemptions, will undoubtedly accelerate the shift in the marketplace. Such regulatory approaches based on the chemical class approach and essential use concept are logistically difficult to implement, but hold great potential and will likely gain wider adoption as the true costs of continued PFAS use come intro greater focus.

10.4 Supply Chain Pressure

Some industry efforts to reduce the use of (certain) PFASs have stemmed from a desire to avoid regulation (*e.g.*, EPA's 2010/2015 PFOA Stewardship Program). However, some industry leaders have demonstrated genuine concern for human health and the environment and have taken prompt action to phase out PFASs from their supply chains upon learning about their potential for adverse impacts. Examples from the latter category include home goods manufacturer and retailer IKEA, which phased out PFASs from all its textile products in 2016,[12] and footwear manufacturer KEEN, which began removing PFASs from its supply chain in 2014 and accomplished its goal in 2018.[13] To take such action, product manufacturers must have complete and accurate information about the chemicals used or present in their products, which is often challenging in and of itself.

Retailers can play a significant role in reducing or eliminating the amount of PFASs that enter the marketplace. Commitments and initiatives from retailers to phase out the use of PFASs put further pressure on manufacturers to pivot from PFASs to safer alternatives. Numerous retail chains in the fast-food, grocery, and big-box store sectors have adopted policies to eliminate or reduce the use of PFASs in their products.[14] Example retailers include Chipotle, Starbucks, Whole Foods, Amazon, The Home Depot, Target, and many others.

Perhaps inspired by these many initiatives in the public, political, legislative, regulatory, and supply chain landscapes, in December 2022, American multinational conglomerate, 3M, announced that it will "exit perfluoroalkyl and poly substance (PFAS) manufacturing and work to discontinue the use of PFAS across its product portfolio by the end of 2025."[15] This voluntary phase out of PFAS manufacturing includes all fluoropolymers, fluorinated fluids, and PFAS-based additive products, as well as the use of PFASs across 3M's product portfolio. According to 3M, this decision is largely based on the increase in regulatory trends aimed at reducing or eliminating PFASs in products, as well as the company's own "changing stakeholder expectations." 3M collects approximately $1.3 billion in annual sales from its PFAS products; however, the company also faces more than 2000 lawsuits over its historic production, use, and environmental releases of PFASs, representing up to $30 billion in potential legal liability for its products.[16] While 3M's voluntary phaseout represents a net positive action for human and environmental health, many other

Figure 10.1 Roadmap for a PFAS-free future.

stakeholders see this as a defensive move to protect the company's finances. Notably, 3M has not made any statement taking responsibility for the harm that may have resulted from exposure to PFASs in its products, nor even acknowledged the links between exposure to PFASs and serious human health concerns.

On the other hand, another major American manufacturer of PFASs, Chemours (a spin-off from DuPont), announced in February 2023 that it plans to continue to invest in the production of fluoropolymers. Current Chemours CEO, Mark Newman, has said that "fluorine chemistries are essential and we believe, based on our technology, can be made responsibly."[17] Despite the recent actions of 3M and the European Union, Newman has stated that Chemours will continue to work with its existing industry customers and "make it very clear to regulatory authorities that this is a chemistry that Europe should embrace."

While these industry commitments differ greatly in their scope and potential impacts, they represent a promising trend in shifting business practices that has arisen from the different sources of pressure outlined in this chapter. The public is more aware of the dangers of PFASs, so they are demanding PFAS-free consumer products, which in turn puts pressure along supply chains. The public is aware of PFASs because of the decades of research by scientists and the growing effectiveness of science communication. Scientific research has shed light on the dangers and ubiquity of PFASs in the environment, and the industry has begun responding to its legal responsibilities in remediating and preventing PFAS pollution. All these factors work synergistically to push the needle toward a PFAS-free future (as shown in Figure 10.1), and the good news is that the momentum of this shift appears to be growing.

10.5 Where to from Here?

Clearly, we believe that a PFAS-free future is possible. We wrote a book about it. However, we also believe that the end does not justify the means. It would go against the principles discussed in this book to replace PFASs just for the sake of achieving a PFAS-free future, without properly considering the hazards and trade-offs of the alternatives. If PFASs are the safest options for certain essential functions, they should continue to be used to fulfill those essential functions until a safer alternative becomes available.

Given the scale at which PFASs are currently used, their alternatives will become widespread as well – as will any adverse impacts of these alternatives. History provides countless examples of seemingly safer alternatives that proved problematic in the long run. Case in point: the first PFAS was synthesized in 1938 while researching new chlorofluorocarbons,[18] which at that time were thought to be safer alternatives to previously used refrigerants. As discussed in Chapter 1, today's problems are often yesterday's solutions.

So how do we ensure that a PFAS-free future does not become another problem that future generations will have to solve? We do not have a magic crystal ball to foretell the future, but drawing from history and our own experience, we offer the following advice for the innovators, manufacturers, and users of current and future PFAS alternatives:

- Narrow the scope to decrease the complexity. Reduce or eliminate unnecessary PFAS uses and seek safer alternatives (nature-inspired, sustainable bio-based, and biodegradable) only where needed.
- Reframe the question and think outside the box. Consider all the different ways in which that function or service could be achieved, including solutions that do not involve chemical substituents.
- Maintain humility and never underestimate uncertainty. We do not know what we do not know. Looking at the world through our limited viewpoints, we can easily miss potential trade-offs and adverse impacts. Start small, test often, and do not hesitate to scale back after you scale up, if new information suggests that would be wise.
- Promote transparency about chemical identities, hazards, and functional uses along the supply chain. Good information is critical to informed decision-making.
- Communicate, coordinate, and share your best practices broadly, including across stakeholder groups. It takes a village to make something as bold as a PFAS-free future a reality.
- Keep a broad focus. Consider trade-offs and synergies with other environmental issues such as climate change and biodiversity loss.

Finally, we hope you will use your failures and your successes on the road to a PFAS-free future to train the next generation of chemists, material scientists, and engineers in the principles of green and sustainable chemistry. A sustainable future can only be built with sustainable chemicals. Now let us work together towards a PFAS-free future.

Acknowledgements

The authors would like to thank André Algazi for helpful comments on the manuscript.

References

1. ECHA, ECHA publishes PFAS restriction proposal, European Chemicals Agency (ECHA), https://echa.europa.eu/sv/-/echa-publishes-pfas-restriction-proposal (accessed 22 February 2023).
2. S. J. Frisbee, A. P. Brooks, A. Maher, P. Flensborg, S. Arnold, T. Fletcher, K. Steenland, A. Shankar, S. S. Knox, C. Pollard, J. A. Halverson, V. M. Vieira, C. Jin, K. M. Leyden and A. M. Ducatman, *Environ. Health Perspect.*, 2009, **117**, 1873–1882.
3. C. A. Moody and J. A. Field, *Environ. Sci. Technol.*, 2000, **34**, 3864–3870.
4. M. Scheringer, X. Trier, I. T. Cousins, P. de Voogt, T. Fletcher, Z. Wang and T. F. Webster, *Chemosphere*, 2014, **114**, 337–339.
5. A. Blum, S. A. Balan, M. Scheringer, X. Trier, G. Goldenman, I. T. Cousins, M. Diamond, T. Fletcher, C. Higgins, A. E. Lindeman, G. Peaslee, P. de Voogt, Z. Wang and R. Weber, *Environ. Health Perspect.*, 2015, **123**, A107–A111.
6. X. C. Hu, D. Q. Andrews, A. B. Lindstrom, T. A. Bruton, L. A. Schaider, P. Grandjean, R. Lohmann, C. C. Carignan, A. Blum, S. A. Balan, C. P. Higgins and E. M. Sunderland, *Environ. Sci. Technol. Lett.*, 2016, **3**, 344–350.
7. S. M. Collins, S.2047 – 117th Congress (2021–2022): No PFAS in Cosmetics Act, https://www.congress.gov/bill/117th-congress/senate-bill/2047/text (accessed 27 January 2022).
8. D. Dingell, H.R.2827 – 116th Congress (2019–2020): Keep Food Containers Safe from PFAS Act of 2019, https://www.congress.gov/bill/116th-congress/house-bill/2827/text (accessed 4 February 2020).
9. J. M. Inhofe, 116th Congress (2019–2020): S.1790 – National Defense Authorization Act for Fiscal Year 2020, Section 329, https://www.congress.gov/bill/116th-congress/senate-bill/1790/text (accessed 5 February 2020).
10. The Real Rob Bilott of ‘Dark Waters’ is Only Getting Started, https://time.com/5737451/dark-waters-true-story-rob-bilott/ (accessed 24 February 2023).
11. PFAS Litigation ‘Second Wave’ Ahead if Biden Delivers New Rules, https://news.bloomberglaw.com/environment-and-energy/pfas-litigation-second-wave-ahead-if-biden-delivers-new-rules (accessed 24 February 2023).
12. A. Blum, *Science*, 2016, **351**, 1117.
13. KEEN Footwear, KEEN Green Paper: The Road to PFAS-Free Footwear, https://cdn.shopify.com/s/files/1/0607/3263/7348/files/KEEN_PFAS-FreeGreenPaper.pdf?v=1677008441 (accessed 24 February 2023).
14. Retailers committing to phase out PFAS as a class in food packaging and products, https://toxicfreefuture.org/mind-the-store/retailers-committing-to-phase-out-pfas-as-a-class-in-food-packaging-and-products/ (accessed 22 February 2023).

15. 3M to Exit PFAS Manufacturing by the End of 2025 – Dec 20, 2022, https://news.3m.com/2022-12-20-3M-to-Exit-PFAS-Manufacturing-by-the-End-of-2025 (accessed 22 February 2023).
16. 3M to Stop Producing PFAS 'Forever Chemicals' by End of 2025 (1), https://news.bloomberglaw.com/environment-and-energy/3m-will-stop-producing-forever-chemical-pfas-by-end-of-2025 (accessed 22 February 2023).
17. Chemours CEO says 'prudent' to keep investing in fluoropolymer plastics|Reuters, https://www.reuters.com/markets/commodities/chemours-ceo-says-prudent-keep-investing-fluoropolymer-plastics-2023-02-10/ (accessed 22 February 2023).
18. Chemours, The History of TeflonTM Fluoropolymers, https://www.teflon.com/en/news-events/history (accessed 22 February 2023).

Subject Index